ADVANCE PRAISE FOR
THROUGH THE GLASS CEILING
TO THE STARS

"Given the chance, I would long ponder trading places with Eileen Collins. Her book with Jonathan Ward is a grand collection of simple, yet sensational moments she experienced— *in Outer Space!*—and in much that led to her getting there. What a *read*!"

—Tom Hanks

"I wrote the song 'Beyond the Sky' and sang it at Cape Canaveral for Eileen Collins's maiden command voyage: 'Once there was a girl with a dream in her heart, wild as the wind was her hope.' This woman with the dream has turned into a serious heroine of the centuries who has taken her place among other men and women in the startling adventure of circling the Earth and leaving it behind. You will love her book: it is exciting, personal, detailed, a good thriller, suspenseful as a Stephen King mystery, and full of hope—that rare quality we all search for. Yeehaw, Commander Collins! What a life you have led and what a tale you have told! Brava!"

—Judy Collins, singer, songwriter, author

"Exciting, motivating, and inspirational are only a few words to describe this story of Eileen Collins and her incredible journey to space. Her tenacity, resilience and persistence come through with each chapter."

—Charlie Bolden, Major General,
United States Marine Corps, retired,
astronaut and NASA Administrator, 2009–2017

"Eileen Collins is a living legend and an inspiration to young people all over the world. We both began our journeys in Elmira, NY, with dreams of doing something extraordinary. Despite being met with challenges at every step, she was tenacious in chasing her ambition of becoming an astronaut—one who would go on to make history as the first female pilot and commander of an American space shuttle. When you come across a story of such determination, it is a reminder that once you set your mind to something, even the sky is not the limit."

—Tommy Hilfiger

"I hope that a young explorer who reads Eileen's book will be inspired to become the first human on Mars. Eileen proves that there is no limit to what we can do if we stay true to our goals and keep moving forward."

—Buzz Aldrin

"Eileen is living proof to youngsters and young ladies that you can do anything you want to do with your life."

—Wally Funk, Mercury 13 pilot

"Eileen Collins and I trained for the shuttle together, flew jets together, and waited together for that first chance to rocket into space. In *Through the Glass Ceiling*, Eileen tells the inspiring story of how she rose through hard work and determination to become a rare exemplar of the 'right stuff,' leading her crews to success in orbit and commanding the first shuttle launch after the *Columbia* disaster. Read, be amazed, then get this book into the hands of young explorers."

—Thomas D. Jones, astronaut and author of *Sky Walking*

"Reading Eileen's story will not only help you get to know this extraordinary woman (mother, wife, survivor, USAF colonel, mathematician, professor, astronaut), but I bet it will help you think about facing life's challenges with a greater sense of strength and determination. She is an inspiration."

—Gwynne Shotwell, president and COO, SpaceX

"This book is endlessly inspiring! Eileen Collins is a trailblazer not only of space but of life. Her story of overcoming adversity to achieve her dream of flying in space will make your spirit soar."

—Andrew Chaikin, author of *A Man on the Moon:*
The Voyages of the Apollo Astronauts

"In the dark days in the aftermath of the *Columbia* tragedy, we looked no further than Eileen Collins to lead us down the long road with the Return to Flight crew to help NASA and the nation believe in ourselves again and resume the human quest to explore. This new book, with the help of master storyteller Jonathan Ward, captures Eileen's thoughtful reflections and compelling story, which serves to inspire us all to reach our greatest potential."

—Sean O'Keefe, NASA Administrator, 2001–2005

"As the commander of first space shuttle flight after the loss of *Columbia*, Eileen Collins lived, worked, and led her crew through a critical phase of history. How she got there is a story as interesting as the events themselves, one that is as inspirational as any in the pantheon of American aerospace heroes. This book takes you inside the life and times of one of NASA's—America's—best."

—Michael Griffin, NASA Administrator, 2005–2009

"Eileen Collins has finally written the book people have been asking for. Growing up on welfare in a fractured family, Eileen's chance to live her own life took form in her long-held love of airplanes; her deep love of flying comes across on every page. Despite numerous obstacles in her path, she turned setbacks into advantages. We're with her in the pilot's seat from page one. This inspiring book takes you into the challenges, the risks, the rewards—the heart of what it took to be first."
—Francis French, space historian; former director of events, Sally Ride Science

"Eileen Collins is an unsung American hero. Her memoir should be 'must reading' for all young girls and boys in school, and indeed for every aspiring student at all levels. She has been a pioneer in everything she tried. The fact that NASA named Eileen as pilot on her first space shuttle flight tells you everything about the trust and confidence she inspires. I have had the honor of working with Eileen in her post-astronaut activities on several boards as well as the White House's National Space Council. That same trust and confidence, combined with just being a natural leader who cares for people, are evident everywhere."
—General Lester L. Lyles, United States Air Force, retired, former Air Force vice chief of staff, and chairman, NASA Advisory Council

THROUGH THE GLASS CEILING TO THE STARS

THROUGH THE GLASS CEILING TO THE STARS

The Story of the First American Woman to Command a Space Mission

Colonel Eileen M. Collins, USAF (Retired), NASA Astronaut

with Jonathan H. Ward

Arcade Publishing • New York

First Edition

Arcade Publishing books may be purchased in bulk at special discounts for sales promotion, corporate gifts, fund-raising, or educational purposes. Special editions can also be created to specifications. For details, contact the Special Sales Department, Arcade Publishing, 307 West 36th Street, 11th Floor, New York, NY 10018 or arcade@skyhorsepublishing.com.

Arcade Publishing® is a registered trademark of Skyhorse Publishing, Inc.®, a Delaware corporation.

Visit our website at www.arcadepub.com.
Visit Eileen Collins's site at eileencollins.com.
Visit Jonathan Ward's site at jonathanward.com.

10 9 8 7 6 5 4 3 2 1

Library of Congress Cataloging-in-Publication Data is available on file.
Library of Congress Control Number: 2021940648

Cover design by Brian Peterson
Cover photograph: Annie Leibovitz / Trunk Archive

ISBN: 978-1-950994-05-2
Ebook ISBN: 978-1-950994-11-3

Printed in the United States of America

To my parents,
James E. Collins and
Rose Marie O'Hara Collins:
Their hardships and resiliency taught me how to face life's
challenges.
—Eileen Collins

To my parents,
John and Patricia Ward,
who prepared me wonderfully but never saw me run this race
—Jonathan Ward

CONTENTS

THROUGH THE
GLASS CEILING
TO THE STARS

PROLOGUE

LEAVING EARTH FOR THE FIRST TIME

It's just after midnight on the East Coast on February 3, 1995. I'm lying on my back, my legs propped above me, in an uncomfortable metal seat. It was tolerable for the first few minutes, but I've been here three hours. An inflatable lumbar support provides some relief to my back, but only for a few minutes at a time.

I should mention that I'm wearing an eighty-pound pressure suit and a helmet, and I'm lying on my parachute pack.

If we launch tonight, I'll become the first woman to fly into space in the pilot's seat of a space shuttle.

It's my first flight as an astronaut, the culmination of a lifelong dream. Women astronauts were not a possibility in 1965, when I first vowed to become one. But here I am, in the cockpit of *Discovery*, on the launchpad at Kennedy Space Center, about to chase down the Russian Mir space station in Earth orbit.

My commander, Jim Wetherbee, and I are running through our preflight checklists. At least I have something to do to take my mind off my aching back.

T minus five minutes. I start the auxiliary power units that will drive the space shuttle's control surfaces. It's beginning to feel real.

T minus two minutes. The orbiter test conductor tells us to close and lock our visors and start oxygen flow to our suits.

T minus thirty-one seconds. The onboard computers take over the launch countdown.

I recall the parting words of our boss when we last saw him this evening. He said, "Remember the Astronaut's Prayer: Dear God, please don't let me screw up!"

Deep breath. Get ready. Cross-check the instruments.

T minus ten, nine, eight, seven . . .

The space shuttle's three main engines ignite, all within a quarter second. I feel the rumble under my seat. On the control panel displays in front of me, the instrument tapes come to life, showing that the engines are spinning up. It sounds like we're in a room on fire or on board a train. The roar grows louder. Still held to the launchpad by huge metal bolts, the entire shuttle stack flexes—the nose pitches down about two meters away from vertical—as those powerful engines try to push us *somewhere*. I focus on the instruments, cross-checking everything as quickly as I can. I'm ready to launch, but I must also be ready instantly for a scrub—a sudden shutdown of the main engines before we light the solid rocket boosters. That's happened several times before in the shuttle program. *Please don't let it happen now.* We won't be going to space today if it does.

Time is almost standing still.

Three, two . . .

The shuttle rocks back toward vertical. All the engines read 100 percent.

One . . .

The solid rocket boosters ignite, brilliantly illuminating our cockpit. Although it's 12:22 a.m., suddenly it's as bright as day outside.

A violent kick in the pants.

Liftoff!

The computer display reads *OPS 102*. The computer is telling us that we have taken off, but there's no mistaking what is happening. Forget a scrub now. We're going flying, no matter what.

Light flashes all around us. It's like being in the middle of a thunderstorm. The sound is that of a blazing inferno. The rumbling has turned to shaking, and *wow!*—I can sure feel the acceleration. Just six seconds after liftoff, we're already going over one hundred miles per hour as we pass the top of the launch tower.

We jerk sideways in our seats as the shuttle automatically executes its roll program, rotating us clockwise. A few seconds later, we jerk to the right as the computer stops the roll. Now we are flying with our heads down. Because of our acceleration, there's no sensation of being upside down, though. I'm pushed back in my seat as hard as if two people my own weight were sitting on my chest.

Wetherbee and I both call over the intercom, "LVLH." Our shorthand for "Local-Vertical-Local-Horizontal" confirms that the space shuttle is no longer using the launchpad as its frame of reference for guidance. It also serves as a communications check. Everything looks good.

We gradually pitch our nose toward the horizon, so that we build up the forward speed we need to reach orbit. The shuttle shakes so hard that if I tried to write something, no one would be able to read it. No wonder the engineers made the computer push buttons so big! If they were any smaller, my shaking hand could easily hit the wrong one.

Less than a minute after launch, we pass through the jet stream, and the rattling peaks. This is much more violent than anything we experienced in the simulators! We're still upside down. I can't see the horizon, because the bright light from the boosters spoils my night vision.

We continue to accelerate, now at two-and-one-half times the force of gravity. We're being hurled into space so fast and so violently that it's almost comical. You just can't believe you can be relentlessly accelerated like this for so long—and we still have another seven minutes to go!

Houston calls, "*Discovery*, go at throttle up."

This is the point where we lost *Challenger* nine years ago. I'm so busy monitoring our instruments that thoughts of *Challenger* never even enter my mind.

The rest of our six-person crew is silent. Wetherbee told us months ago that he did not want us to yell, "Yeehaw!" during the launch. He wants us focused on our tasks. We must minimize the intercom chatter in case something goes wrong and we need to take immediate action. I feel fortunate to have the consoles and controls to watch, and that I can steal an occasional glance out of the windows. The crew members in the mid-deck feel the shaking and hear the noise, but without windows, all they can see are the storage lockers.

Two minutes into the flight, our solid rocket boosters begin to burn out. We see *PC < 50* on the display, meaning that the chamber pressure inside the boosters is decreasing. There's a tremendous *pop!* as explosives fire to separate the cylindrical boosters away from the side of the orbiter and its fuel tank. Thank goodness they are gone. If anything went wrong during our launch, we could do nothing as long as the boosters were firing. There is no way to shut off the solid rockets until they burn out on their own.

The computer alerts us with *OPS 103*. It has moved on to the next part of the automatic flight control program. We're powered by the shuttle's main engines alone, which burn liquid hydrogen and liquid oxygen. The ride feels much smoother, almost like being in an airplane. The autopilot guides our flight, but Wetherbee or I could take control and fly manually if required. However, it could end our primary mission to rendezvous with the Russian Mir space station, in orbit 223 miles above Earth. The computers are programmed to fly *Discovery* on autopilot during ascent, because the flight envelope is so narrow. A human could hand-fly the ascent safely, but not accurately enough to hit the narrow corridor needed for a rendezvous.

We had briefly dropped to one g, but now we are accelerating again. Wetherbee calls out, "Two-engine Ben." That means that if we lose one of the three engines, we can safely fly across the Atlantic to the emergency landing field at Ben Guerir, Morocco. Shortly after that, he calls out "Negative return." We can no longer fly back to Kennedy Space Center if there's a problem.

We enjoy a smooth and quiet ride for the next six minutes. As *Discovery* burns off fuel and oxidizer, the shuttle's weight decreases, so the constant force from the engines accelerates us faster and faster. Soon we have hit three g's, meaning that if I weighed 150 pounds on Earth, now I'd weigh 450 pounds if I could stand on a scale. It's uncomfortable but not intolerable. Reaching for switches is tricky. My arm feels so heavy that I need to concentrate to avoid hitting the wrong one.

Wetherbee calls, "Three g throttling." The flight computer throttles back the engines so that we don't exceed the maximum safe loads on the attachments between the orbiter and fuel tank.

Time clicks by quickly. We call, "Fine count." The software zeroes in on the precise microsecond and point in space where it will shut down all three engines. An early shutdown might render us too slow to catch up with Mir. Several seconds too late might result in the fuel running out, causing the engines to cavitate—kind of a burp, only a lot worse . . . one that results in a *kaboom*.

Wetherbee and I are watching the "bug," a triangular marker crawling slowly along a time line on the computer's display. Ideally, when the bug crosses the appropriate tick mark on the time line, we will feel MECO (pronounced MEE-co). That's an acronym for main engine cutoff, one of the most important milestones in our flight. Wetherbee and I have our hands on the controls, ready to shut down the engines manually if the computer fails to do so.

Here it comes . . .

MECO.

Silence.

There is no jerk forward. We haven't stopped; we are just no longer accelerating. We instantly transition from three g's to weightlessness. Everything slowly starts floating about the cockpit. My helmet ring rides up around my face. It's hard for me to see my checklist. I'm strapped in very tight, but my spine stretches, since it no longer has to support the weight of my head and torso. I take a deep breath. My checklist pages fan open. There's dust in the air around me. Tethers flap in slow motion. It's surreal.

I snap back into focus. *Quick, jump on the checklist.* As pilot, I must shut down the three auxiliary power units, throw switches to tie two of the shuttle's electrical buses together, and purge and vent the main propulsion lines. If I don't perform these critical actions quickly and correctly, we're in big trouble. These steps will be automated in future spacecraft, but for the moment, the mission depends on me. I concentrate on not making a mistake.

Focus.

Current and future women pilots are counting on me to do a perfect job up here. I have just become the first woman to pilot a space shuttle, and the mission is just barely underway. We've had just eight minutes of flight so far. Eight more *days* of intense work await me. The quality of my performance will set a standard for those who follow me.

FOCUS.

As I run through my checklists, our flight engineer, Mike Foale, sitting behind me, says, "Eileen, stop working so hard and look out the window. This is your first sunrise from space!"

I look up and see a beautiful and strange rainbow wrapping around the edge of our planet, separating the blackness of the Earth's night side from the blackness of outer space above it. I've seen photos of this before, but this is different. The colors are bold: yellow, orange, red, blue. I've never seen or imagined anything so beautiful. I'm floating and breathing carefully, still unwinding from that crazy and intense ride to orbit. For now, my logical brain has to take a back-seat to the emotional impact of what I'm witnessing out the window.

Wow, the EARTH IS ROUND!

Of course, the Earth is round!

But as a pilot who has looked out the front windows of cockpits for decades, this is my first time to actually see the curvature of the Earth.

Yes, the Earth is round.

And after a lifetime of striving to be here, I am finally in space.

Chapter I

ELMIRA

I was born Eileen Marie Collins on November 19, 1956, in Elmira, which straddles the Chemung River in upstate New York. My father was James Edward Collins, and my mother was Rose Marie O'Hara. My brother Edward (Eddie) is thirteen months older than me. My sister, Margaret (we call her Margy), is fourteen months younger. My brother James (everyone else calls him Jim, but I call him Jay) was born in 1962, after my mother suffered several miscarriages and a stillborn baby.

Elmira was a busy stop on the Underground Railroad before the Civil War and the site of an infamous prison camp for captured Confederate soldiers. Our town produced its share of famous citizens: film producer Hal Roach (who made the Laurel and Hardy movies and the Little Rascals short films), fashion designer Tommy Hilfiger, and Ernie Davis (the first African American to win the Heisman Trophy). Mark Twain's wife hailed from the town, and Twain spent the last years of his life there.

In the 1960s, Elmira's population peaked at just over 45,000. Unfortunately, like many other small towns in New York, its population has gradually declined since then. But it was a great place to grow up, with plenty of parks and woods easily accessible by a kid on a bike or walking. Our family still owns my grandmother's house. I regularly go back to visit my high school friends and the parks and historical sites I loved as a child.

I might have spent my whole life there, perhaps teaching math and science, if things had turned out differently.

My Father

That's not to say my childhood was idyllic. Far from it. My father was an alcoholic—the worst kind—and our family paid the consequences of his disease. It was awful to live through, but it shaped me into who I am today.

Dad's ancestors came to the United States from County Cork, Ireland, in the mid-1800s. They owned a farm in Ridgebury, Pennsylvania, and moved to Elmira in about 1920 to establish a pub on Davis Street. Dad was born on July 4, 1926. He enlisted in the Navy near the end of World War II, serving in the Pacific. Soon after the war ended, his mother wrote to the Navy and asked that he be discharged, because his father was dying. The Navy released him, and he came home to spend the next nine years running the family pub. When he married my mother at age twenty-seven, she insisted that he quit working at the bar. He became a surveyor for the city.

I loved hearing about his work. He brought home his papers, maps, pens, protractors, transit, and other equipment. He showed me his detailed maps and plans for the roads and building sites he was surveying. Exposure to all this fascinating material and the techniques of surveying led naturally to my interest in mathematics, maps, and eventually flying.

Dad was intimidating, loud, and highly opinionated. He reminded me of Jackie Gleason's character in *The Honeymooners*, except that Dad was much smarter. He was tough but fair. He knew right from wrong, and he didn't hesitate to call people out for doing the wrong thing. He was a strict disciplinarian. He would not tolerate "weak" talk from his kids. He abhorred indecisiveness and laziness. We didn't want to set him off, because he could really lay into us for not living up to his standards.

That was my father's *good* side.

But then there was *the Place*. As far as I know, no one actually called it by its real name, Collins' Restaurant and Pub. Everyone just referred to it as the Place. Although he no longer worked there, Dad made us go in with him on weekend afternoons.

Everything about the Place scared me. It was dark, the high-backed booths were foreboding, there was a bar with liquor everywhere, and the air was heavy with smoke. When Dad drank, it terrified me. My siblings and I would sit in a booth while waiting for Dad to finish drinking at the bar. It was worse than boring, because there was an undercurrent of dread about what was going to happen next. We knew that Dad's personality would change after just one drink.

It was such a waste, because when he was sober, he was so wise. He could read people better than anyone I've ever met, probably from so many years tending bar. However, his drinking was completely out of control during my childhood. He might be sober for two months, then drunk for two weeks straight. You could never predict how long either stretch would last.

We sometimes appeared to be the perfect family—going to church, eating brunch at Grandma's house, then taking a trip to the cemetery to visit dead relatives. (I did enjoy running around there!) But other times, without warning, Dad would come home roaring drunk late at night. My mother would yell at him, he would throw chairs and curse at her, and eventually she'd kick him out. She often just locked him out if he wasn't home by nine o'clock. Once, he pounded on the locked door so hard that the glass shattered. More typically, he'd bang on the door and yell and curse. We four kids would be upstairs, hiding under our beds, scared to death. After a while, he would give up and walk over to Grandma's house.

He had several DUI convictions. He ended up in jail or the hospital many times after his long drinking spells. We could stop being afraid and take a little breather while he was gone.

Dad's stints in rehab wore us down financially, because he wasn't able to work. We moved from a two-bedroom house into subsidized housing in "the projects" when I was seven. My mother complained about it the entire time we lived there. For us kids, though, it was a good location, right on the edge of town next to Hoffman Creek. There were plenty of outdoor spaces for us to explore. We built forts in the woods with the neighborhood kids.

Money became tighter and tighter. We were on and off welfare and dependent on food stamps. I remember separating paper products from food in the grocery store checkout line, as food stamps would not pay for paper products. We pinched pennies, relied on coupons, and rarely bought name-brand items. We ate out maybe twice a year. On the rare occasions when we could get a burger, we couldn't have cheese on it, because cheese cost an extra five cents. Pizza for school lunch with my friends? No way. I brought bologna sandwiches from home. Thank goodness the family was able to get financial assistance to allow us kids to attend Catholic school and summer camp.

Dad's drinking became so uncontrollable that when I was nine, my mother threw him out for good. He never lived with us again, although my parents remained legally married the rest of their lives, and he continued being our dad when he was not drinking.

When Dad moved out, I felt the world had ended. I had never heard of parents splitting up. It just wasn't done back then. Mom took a job as a stenographer at the Elmira Correctional Facility. I was the only person I knew whose mother had to work. I felt inferior and disgraced.

Dad once told me, "There's something about this town—too many ghosts here from my past." (That might have been more than a figure of speech! Dad told us about the Irish wakes at his parents' house when he was young, with dead relatives in caskets in the front room.) His doctor warned him, "You are never going to stop drinking unless you get out of Elmira." Dad moved to Rochester and worked at the post office in the suburb of Henrietta. He took better care of himself during the weekdays when he was in Rochester, but then he'd come back to Elmira on the weekends and drink. He stayed at Grandma's house when he came to town, because Mom refused to allow him back home.

Mom always told us that Dad would never get better until he recognized that he had a problem. That of course is the first step of the Alcoholics Anonymous Twelve-Step Program. AA eventually helped Dad tremendously, and I highly recommend the program to anyone who struggles with an addiction. Although he never fully recovered, he made good friends, became a speaker, and helped many other people recover. He understood all too well what they were going through. He suffered more than his share of relapses throughout the rest of his life.

I became closer to Dad when I was a teenager, and I learned quite a bit about how adult men think. He was always honest with me and admitted his weaknesses, although he had excuses to explain away his alcoholism—experiences and family history being among them. He talked about it philosophically. And yet, he still kept drinking.

Young Eileen

I took speech therapy in second grade because I stuttered badly. I was painfully shy, afraid to speak up at school or in the presence of my friends, even though I wanted to. That was doubly hard on me, because I'm an extrovert and enjoy being with people. I hated that I was afraid to speak up or say anything out loud. I blamed my shyness on being the youngest student in my class. Fortunately, I had no trouble talking with my family.

My parents were huge proponents of Catholic school. Dad didn't have anything against public schools, but he wanted us to have a religious education. I attended Catholic school through tenth grade.

True to their reputation, the nuns strictly enforced discipline—and they were also excellent teachers, smart, and loving. At St. Patrick's, my school until eighth grade, we learned how to handle our interpersonal differences on the playground without adult supervision or intervention. The philosophy was: children learn best when there is law and order in the classroom and enough freedom to explore and learn on their own.

With the Cold War in full swing during my first few years of school, we feared an imminent nuclear attack from the Soviet Union. Civil Defense drills were part of our routine. The sisters lined us up and marched us quickly and silently down into the school's basement. There, we'd squat down with our heads to the wall and backs toward the hallway and our hands over our eyes. We never understood what we were doing or why. We just did as we were told. Thank goodness I was too young to have comprehended the horror of a nuclear attack!

Growing up during the Cold War didn't affect me as much as being a child of the Space Age, born a year before the Soviet Union launched Sputnik. Though too young to remember Project Mercury and America's first ventures into spaceflight, I became aware of astronauts and space travel by watching *Buck Rogers* reruns on TV.

By fourth grade in 1965, space had captured my imagination. I read an article about the Gemini astronauts in *Junior Scholastic* magazine. They were my heroes! I wanted to be just like them. Although there weren't any women astronauts, that didn't deter me. I decided that I was going to be a lady astronaut anyway. As my backup plan, if I couldn't be an astronaut, I would marry one.

I loved math and science. I always maxed out on the aptitude tests for math, physics, mechanics, shapes, and logic. However, I scored relatively poorly on the verbal tests. I never tried to excel in academics, even though I enjoyed them. The boys ridiculed smart girls, so why on earth would a girl want to be smart?

I was content to be a B student. I did the amount of work needed to get by—nothing more. I was not competitive. I didn't want to stand out and was much more interested in fitting in socially. Rather than pushing myself to excel academically, I joined the home economics cooking club. That drove my dad crazy. He constantly chided me: "Don't follow the crowd! Think for yourself! If all your friends jumped off the Walnut Street bridge, would *you* jump off the Walnut Street bridge?" Then he'd cap it off with "What does *Eileen* want to do?"

Mom offered similar advice, which sounded more comforting coming from her. "There is no one else in the world exactly like you! Every person is an individual with their own different characteristics. We are here on this planet to fulfill a need somewhere in the world."

Feeding My Imagination

Throughout my childhood, my mother took us four kids to the Steele Memorial Library on Church Street. The endless rows of tall bookshelves initially intimidated me, but soon the library became one of my favorite places. I always came home with armloads of books. My initial interests were dogs, horses, and bugs. As I grew older, I was captivated by any story of a faraway land or travel to unusual places.

By eighth and ninth grade, I chanced upon the library's aircraft section. The more I learned about planes, the more my curiosity kept me coming back. I couldn't read enough about flying. I studied how airplanes fly. I read everything I could about civilian pilot explorers like Charles Lindbergh, Amelia Earhart, Jackie Cochran, and Nancy Love. I idolized the daring military aviators from the world wars. My mom sowed the seeds for my lifelong passion for flight just by taking me to the library.

I attended summer camp for seven weeks annually from age seven to twelve. Camp was a lifesaver for me and my siblings—a safe place where we could just be ourselves. It gave my mom some time to herself and enabled her to work without worrying about what we were up to. Camp El-Ne-Ho was sponsored by the Elmira Neighborhood House, a racially diverse community action organization created to help foster positive change for individuals and families.

A perpetually happy old man named Charlie Kromer came by camp every day to watch us play and revel in our excitement. I learned that he was a benefactor and director of the organization. Mr. Kromer became a role model for me later in life, because I now appreciate the difference one caring adult can make in a child's life and how summer camp can inspire any kid.

I attended El-Ne-Ho day camp at Sayre Wells Woods. We sang traditional spirituals and corny camp songs every day, up and back, during those rides on the rickety bus. After five weeks, we moved to overnight camp at Harris Hill, where we then spent two weeks living in cabins.

I have no doubt that my summer camp experience enabled me to succeed in the military. I learned how to swim, lifeguard, run, play sports, clean the cabin to inspection standards, get along with kids of different backgrounds, plan and execute projects, do woodworking, shoot bow and arrow, not complain about food, keep going even when tired, and honor our country's flag. I felt free, competent, and safe at camp.

Harris Hill at the time was the "Soaring Capital of the World." (Even today, it is home to the National Soaring Museum.) At camp, I'd watch gliders circle overhead on the summer air currents— silently and effortlessly, like birds. My dad loved planes, even though he'd been a sailor in the Pacific during the war. He often took us to the glider field or the local airport on weekends. We sat on the hood of his car and watched the planes and gliders take off and land. I yearned to know what it was like inside one of those planes, but we didn't have the money for flying lessons.

Maybe someday.

If Mom planted the seeds of a love of flying, Dad helped them grow into a possibility. My brother Eddie and I caught the flying bug from our dad. Eddie talked about joining the Air Force someday. His interest rubbed off on me, too, but I didn't talk about it at the time. Eddie loved to build Estes model rockets, and I always accompanied him when he launched them behind the high school.

We moved out of subsidized housing and into a small house on West Second Street when I was thirteen and my brother Jay was seven. Despite the big age difference between us, I enjoyed hanging out with him. We'd pitch baseballs or explore the neighborhood. Temperamentally, he and I were very much alike. We played by the rules and tried to be helpful. I was so proud when Jay won the Little League's Best Sportsman Award for the entire city!

Growing Up Fast

In June 1972, the remnants of Hurricane Agnes moved up the East Coast. The rain started in Elmira while I was in school taking my biology final at Notre Dame High School on Wednesday, June 21. I rode my bike home—three miles—in the pouring rain. The hurricane stalled over Pennsylvania and Upstate New York. The heavy downpour continued all day Thursday and Friday and into early Saturday. The rain never let up.

Mom woke us up on Saturday morning to tell us that the dam downtown was overflowing. I walked down Columbia Street and saw an unimaginable sight: a tremendous flood of dirty brown water pouring over the dam. For a fifteen-year-old girl, it was pretty exciting! I spent the morning in fascination, watching the river rise.

As the river overflowed its banks, the water slowly crept through the city toward our house, five blocks from the dam. Now it wasn't exciting anymore; it was dangerous. We had to evacuate.

Mom took us to the evacuation center at Booth School, where we spent the afternoon in the second-floor library with a couple other families. Just before dark, an amphibious vehicle took us farther away from the flood. We spent the next three days and nights in the basement of St. Casimir's school with about twenty other families.

When we finally went home, our house was the most disgusting mess imaginable. Although the waters had receded, mud and trash covered everything. The smell was overpowering. The water crested at about two inches deep within the first floor of the house, seeped up into the walls, and mildewed the furniture. We had to throw out everything we couldn't clean. We had no running water for a week and no electricity for three weeks.

I volunteered to help out with the relief efforts at Arnot Ogden Hospital. I helped patients in physical therapy and the X-ray department, and I also worked at the snack bar. I didn't make any money, but it was so wonderful to feel *needed*. For me, something positive came out of the Great Flood of 1972. I felt I could make a difference, and that was a powerful motivator for a young girl.

That summer, between my sophomore and junior years, I realized that I needed to switch schools to continue to improve my self-esteem. While Notre Dame High School was excellent academically, I never felt like I fit in. I felt excluded from the cliques, social life, and other activities. My string of disappointments began in second grade, when I couldn't get into choir, even after my mother made me re-audition. I was bullied by other kids at school, even some whom I had once considered to be my friends.

In those days, competitive sports were not an option for girls. Everyone felt that cheerleading was the be-all and end-all for any worthwhile girl. I tried out for the cheerleading squad in sixth, eighth, ninth, and tenth grades, and I never made the cut. I was slightly overweight, and I worried that the coaches didn't like my appearance. In tenth grade, I decided to audition for the lead role in *My Fair Lady*. I practiced for a month. However, I talked myself out of auditioning on the big day. *You know what? I really can't sing*.

A fresh start in a new school seemed to be the best option. Some of my friends from my Catholic primary school already attended the public high school, Elmira Free Academy, which would ease my transition.

I convinced my parents to let me switch. This was hard news for them to hear at first, since they believed so firmly in the benefits of Catholic schools. The financial argument turned out to be the most compelling. The burden of having three children in Catholic high school in the upcoming year would be too great, especially with our house in shambles after the flood. That summer, Eddie, Margy, and I all transferred to public school. Margy and I also took our first paying jobs—counting the collection money for Father Joseph Egan at St. Patrick's Church.

My first year at my new high school went well, but things were getting very hard for my mom. She began to struggle emotionally after the flood, when I was about fifteen years old.

First, I should say that she was the best mother I ever could have asked for. She was honest, wise, a great cook, and she loved her children deeply. She took us to the Watkins Glen State Park pool every summer weekend, just to get us out of the house.

Mom must have had the explorer gene in her, because she took us on several vacations when my dad was either working or missing in action. She wasn't afraid to venture out to new places, just herself and her four young children—state parks, Niagara Falls, or Ocean City, New Jersey. As a huge extrovert, Mom enjoyed being with people and attending social events. She had a close group of friends from Al-Anon, the spousal support arm of AA. They'd sit around our kitchen table for hours, smoking and talking.

However, Dad's drinking and then the aftermath of the flood on our house and possessions became too much for her to deal with. Eddie graduated from high school in June 1973 and left home to attend school in Minnesota. That took a toll on Mom, too, since Eddie was the unspoken father figure, with Dad gone. As our financial problems continued, Mom always worried about her next paycheck.

Her deterioration was particularly noticeable by the start of my senior year of high school. She would get angry for no apparent reason. She yelled at people she didn't know. One time she screamed at a man for taking "her" parking spot, and I honestly thought he would kill her.

One Saturday morning in January 1974, she finally cracked. Mom seemed overwhelmed by her problems. Sobbing uncontrollably, she told me she had no reason to live. I was so worried about the extremes of her emotions that I called the emergency room. A woman questioned me for a few minutes, then said there was nothing she could do and hung up. But my mom was clearly suicidal. About half an hour later, she came down the stairs yelling incoherently, swallowed an entire bottle of blood pressure and thyroid pills, and went back upstairs crying. I immediately called for an ambulance.

I followed the ambulance to the hospital. It was the first time I ever drove a car alone, because I had just earned my driver's license the day before.

On Monday morning, they institutionalized Mom in Binghamton, New York, for four weeks of observation and treatment. My grandmother came over and spent that night on our couch. She couldn't drive and was far too old to take care of a houseful of youngsters. I told her we could get along fine on our own.

There was no one else available to watch me and my two younger siblings, so we lived by ourselves without an adult in the house for the next four weeks. As the oldest person at home, I was in charge of the family. I cooked, drove, shopped, shoveled snow, and cared for Margy and Jay. My cooking wasn't exactly up to adult standards, though, and my menu selections consisted of Chef Boyardee pizza, Hamburger Helper, and Spanish rice.

On weekends, we visited Mom in Binghamton. She was so drugged up that she was unaware of her surroundings. She probably didn't even know we were there. That big old ugly institutional facility terrified me, and I didn't want my mother in there. Fortunately, a new facility opened in Elmira for day use by mentally ill patients. They admitted Mom in February.

The doctors corrected Mom's thyroid medication and adjusted or discontinued the other pills she was taking. She eventually got better. In June, she was well enough to attend my high school graduation.

What Does Eileen Want to Do?

That winter and spring, I realized that I needed to actively take charge of my life. I couldn't just live passively and let things run their course.

I saw firsthand what could happen to two wonderful and loving parents when they let bad choices ruin their lives. That spring, I swore I would never let it happen to me.

I learned how important it is for a person—and their family—to be in charge of their own medical situation. Take the right medications, don't take pills you don't need, don't smoke (my mother chain-smoked from age ten onward), don't drink to excess, and keep physically fit.

I became a runner, because I didn't want to end up in bad physical shape like my mother. I saw what damage Dad was doing to himself from literally falling down drunk. I was afraid Dad would seriously hurt himself, and I was afraid my friends would see him drunk. I didn't want to live the rest of my life in fear—both for him, and of the same thing happening to me.

As a consequence, I didn't smoke, rarely drank alcohol, and never took unprescribed pills or illegal drugs. I wasn't about to allow alcohol to control me like it did my father.

I vowed I would never get married. I couldn't go through domestic hell like this again. I told myself I was going to "marry an airplane."

I knew I was in a toxic environment and realized I had to get out of Elmira to escape the fear and uncertainty I was living in. The structured life of the military seemed to be what I desperately needed in order to keep my sanity. And I wouldn't need any money to enlist.

So one weekend when Dad was visiting, I told him that I wanted to join the Air Force. He was livid. The vehemence of his reaction surprised me. "You are *not* joining the Air Force!" He objected so emphatically that I dared not argue with him.

I dropped the idea for a few weeks at least but couldn't put it out of my mind. Finally, I decided I would create my own future. I called the town's Air Force recruiter and set up an appointment for four o'clock one afternoon.

I drove over to the office. The door was locked, and no one was in sight. After hanging around for a few minutes, I drove home, feeling crushed. I couldn't understand why the recruiter hadn't shown up. *Why didn't the Air Force want me? Was it because I was a girl?* After that, I completely gave up dreaming about joining the Air Force.

That recruiter actually did me a *huge* favor by not showing up. At that time, I had no idea about the difference between an enlisted person and an officer. I didn't know that you had to be an officer to fly, and that the only way to become an officer was to have a four-year college degree. Had I enlisted right out of high school, I might never have flown an airplane. It was only later that I realized I'd dodged a bullet.

I neared the end of my senior year of high school feeling frustrated and without any compelling sense of direction. I never thought I could do big things; I just figured I would do *something*. But what?

My wake-up call came at the awards ceremony for our graduation. I watched my classmates walk the stage, winning awards and accepting scholarships. Then the overpowering question hit me: *What had I done in high school? I had gotten by.*

While I'd entertained thoughts of becoming a math or science teacher, I didn't take any math or science classes my senior year. Instead, my senior courses were blow-off subjects like film studies, speed-reading, and The Law and You. I hadn't competed or tried hard for anything. I put far too much emphasis on my social life, which I thought much more important.

Suddenly, I felt as though I'd completely wasted those precious years of my life.

My advice for young people is to make the most of your high school years. Don't coast. Don't talk yourself into low self-esteem by constantly comparing yourself to everyone else's best qualities. Challenge yourself. Take as many classes as you can, because they're *free*. Get as far ahead as you can in high school, so you don't have to pay for those courses in college. Use high school as a chance to learn about yourself in what is actually a relatively low-risk environment, although it certainly feels high-risk in the moment!

I wish I had paid attention to my parents' advice earlier. I guess I just wasn't ready to *hear* what they were telling me.

That very day, and from that moment on, I decided to push myself to be the best I could possibly be. I was finally ready to ask myself the question my father would always ask me: "What does *Eileen* want to do?"

Chapter 2

COLLEGE AND FIRST FLIGHT

Having decided to attend college, the question became *Where?* Although I applied to several schools, I knew I needed to stay close to home to care for my mother and siblings. Corning Community College was the logical choice, as I could live at home and drive half an hour to classes.

Corning was a lifesaver. My experience turned me into a *huge* proponent of community colleges. I saved money by living at home and driving to class daily. Best of all, the college had the program that I wanted, an associate degree in math and science.

My friends tried to convince me that community college was just like high school. It most certainly was not. It was college, through and through. I had plenty of challenging work, the professors were stellar, and I had individual responsibility for what I learned.

Perhaps most important, I had to pay for it myself —*real* money that I had to earn. I worked thirty hours a week. My evening and weekend part-time jobs were not the most glamorous in the world—maintaining the grounds and selling tickets at a putt-putt golf course, serving as hostess and salesperson at the Century Housewares catalog showroom, or making sandwiches and working the counter at Pudgie's Pizza.

High school was free (at least for me), and I took it for granted. Now that I had to pay for every class with part-time jobs, loans, and a few grants, I took my studies much more seriously. I matured a lot in those two years.

I took eighteen or nineteen credit hours per semester. I concentrated on math, taking two or more courses per semester to catch up. The rest of my studies were science and core requirements.

I enjoyed math, even though I couldn't always internalize the concepts immediately and I never fully understood calculus the first time I heard it in the classroom. At home, I studied every free minute. I memorized concepts and worked problems over and over until I felt I'd covered everything.

My aim was to be a straight-A student. Although I didn't quite meet that objective, I never gave up. I gave it my full effort and graduated in four semesters, earning seventy-four hours of credit, even though only sixty were required for the associate degree. And I impressed myself with my results, considering my lackluster performance in high school.

Dreams of Flight

When I wasn't studying, working, or caring for my family and house, I read everything I could about flying. Some of my favorite books included *Fate Is the Hunter*, *The Stars at Noon*, and *God Is My Copilot*. I subscribed to *Air Force Magazine* and read it from cover to cover. I read a nonfiction book on the design and procurement of the C-5 Galaxy transport plane, with all the mistakes and lessons the Air Force learned along the way.

My favorite reading topic was military history—books like *Into the Mouth of the Cat* (which told the story of pilot Lance Sijan and how he evaded capture in Vietnam) and a novel about military leadership called *Once an Eagle*. Books were by far the most important influence upon my decision to become a pilot.

I became particularly interested in the Air Force's "Century Series" aircraft, high-performance fighter jets designed in the 1950s and early 1960s. These included the F-100 Super Sabre, F-101 Voodoo, F-102 Delta Dagger, F-104 Starfighter, F-105 Thunderchief, and F-106 Delta Dart. With features designed for their unique military missions, each aircraft represented huge leaps in aeronautical technology, with advanced engines, swept or delta wings, and sophisticated avionics (flight systems). The Century planes were all supersonic, and their pilots could fly farther, faster, and higher than anyone before.

I could easily imagine myself above the clouds, high up and far away from all the problems on planet Earth, looking down upon cities or seeing distant oceans and polar ice caps. The sky would be a very dark blue at such high altitudes. The air would be cold and thin, and the horizon hundreds of miles away. In a single-seat jet like that, I would be alone in the world, like a Greek god flying over the planet or an angel playing in the clouds, going anywhere I chose.

I couldn't think of anything I would rather do than fly.

But why *me*? I wondered why I was apparently the only young woman in my town with such an unbridled passion for aviation. Perhaps I'd inherited it, or maybe my parents had nurtured my interest before I was even aware. Whatever the reason, I was convinced that flying represented the best possible thing I could aspire to.

I researched the Air Force's Reserve Officer Training Corps (ROTC) program. Cornell University and Syracuse University were Upstate New York's two schools offering ROTC. I visited both campuses during my second autumn at Corning. I had difficulty connecting with the ROTC people at Cornell, but Syracuse answered the phone. I spoke with Sergeant Ault, set up an interview for January 1976, and submitted my application.

Now I had to tell Dad about my plans. After he forbade me to apply to the Air Force two years earlier, I was convinced he would be furious about me taking this step. But I told him about the ROTC program. I explained that they would pay for my tuition, books, and fees, and that I would have a guaranteed job after graduation.

He listened to me without responding—neither forbidding me nor encouraging me. That was actually his way of letting me know he was okay with my decision. His silence was tacit approval. That was a major victory, as far as I was concerned.

Syracuse notified me in March of my acceptance into the ROTC program. I was one step closer to the Air Force.

As I was completing my studies at Corning in 1976, the Air Force announced a new test program to train women as pilots. Women hadn't been permitted to fly military planes since the Women's Airforce Service Pilots (WASPs) in World War II. WASPs were civilian pilots, so in effect women had never actually been military pilots. The Navy began accepting women for pilot training in 1974, and now the Air Force was catching up.

The Air Force's senior leaders were unsure whether women could deal with the difficult lifestyle and the demanding physical strength required of pilots. They worried about the very real consequences of the enemy capturing women pilots during combat. During the Vietnam War, which had just recently ended, the enemy captured and tortured a significant number of Air Force pilots and Navy aviators as prisoners of war. No one wanted to think about what would happen to women taken prisoner, or what their male colleagues might do to keep their female counterparts from being tortured.

So, the Air Force decided to accept women as pilots but only for noncombat missions. Women would be able to fly trainers, cargo planes, tankers for midair refueling, and administrative flights. Fighters, bombers, close air support, reconnaissance, and any other combat planes were off-limits.

The first group of ten women student pilots were chosen for this test program from active-duty personnel—commissioned officers with a four-year college degree and a minimum time already served. It was an extremely competitive selection process. When the ten women were announced, I noted their names and followed their progress as closely as I could. I buckled down on my studies even more intensely, hoping that I might be able to follow in their footsteps someday.

Basic Training

Shortly after graduating from Corning, I reported for six weeks of basic training. I got a haircut a few days before I left home. The stylist clearly understood about basic training, because he cut my hair so short that I—seriously—looked like a guy.

I flew to Columbus, Ohio, and took a bus to Rickenbacker Air Force Base. I disembarked in front of the barracks, a World War II–era three-story building. Just then, a tall, pretty woman with long blond hair and two suitcases in her hands seemed to be in a hurry to get onto the bus. I asked, "Where are you going?" She said, "I'm joining the Navy!"

What was that all about?

Perhaps she didn't like the accommodations. The women bunked on the third floor; the men had the first two floors. The rooms were barren, with tile floors and bare walls. Each room housed three occupants, who each were assigned an old cot and one aluminum wardrobe unit. The showers were open bays. I figured that the girl leaving in such a hurry must have taken one look at the facilities and decided that this wasn't what she had signed up for.

My two roommates were already at each other's throats. They fought about where to place the few pieces of furniture. They became increasingly bitter and nasty, shouting and name-calling. I couldn't figure out what was such a big deal on our very first day.

Somehow, I calmed them down. I reminded them that we had to live with one another for six weeks. One of them told me, "Would you tell that person over there that I'm not speaking to her?" And then the other said to me, "Well, you can tell that person over there that I'm not speaking to *her!*" I couldn't believe I was seeing two adult women—who hoped to become Air Force officers—exhibit this level of pettiness.

The next day, which was our first day of training, the captain called me into his office. I saluted. "Miss Collins reporting as ordered, sir."

He said, "I heard about what happened in your room. Miss Collins, I am going to hold you personally responsible for everything that happens in that room. You keep the peace in there. *You* are responsible."

I saluted and said, "Yes, sir."

Not only did I have to deal with these two women, whom I seriously doubted were officer material—now *I* was at risk if they couldn't control themselves. They never actually fought again, thank goodness, and I was able to get along with them individually. I never felt close to either one of them, though. At least we kept it civil for the rest of basic training.

Neither one said a word directly to the other for six solid weeks. Everything went through me.

"Tell that person to clean up their stuff."

"Tell that person to close the window."

That was my introduction to military life.

Nonetheless, I absolutely loved basic training. I jumped right into the routine: running every morning, marching every afternoon, daily lessons in the classroom. In 1976 (unlike today), women had different fitness standards than their male counterparts. For example, men had to run one and one-half miles in under twelve minutes, but women only had to accomplish a twelve-minute "continuous run." The men ran on the roads around the base and competed with one another. The women just ran around a gravel track across the road from our barracks. Some women just walked the track. No one monitored or coached us.

I didn't think it was fair that women got a break from physical requirements. After the first day, I asked the captain, our flight leader, if I could run with the guys. He agreed. I ran with the men from the next day onward. I wanted to improve. Competing with the guys was a great motivator!

I found marching to be similar to cheerleading. There was a choreographed aspect to it, with flanking and column movements to memorize. I enjoyed learning and executing the routines as precisely as I could.

Academics was my favorite part of the program. Our courses included Army and Air Force history, development of the airplane and theory of flight, organization of the Air Force, individual outstanding leaders, and strategic and tactical military missions. I particularly enjoyed stories about pilots. I still have all my workbooks from that summer. Of all the awards I've received over the years, I particularly cherish earning the top academic award for our entire basic training class.

The icing on the cake that summer was the field trips. We visited the flight line, and I sat in the cockpit of an A-7 Corsair II. It was my first time inside the cockpit of an airplane. It smelled of JP-4 fuel, electronics, and dirt. There were switches and circuit breakers everywhere. The maintenance chief eyed me and my movements carefully, probably checking to ensure that the ejection seat was pinned so I wouldn't blow myself sky-high! *Me! In the pilot's seat of an airplane that probably saw action in Vietnam.* I could hardly believe it. *Will I ever be able to fly a plane like this?*

Our training included a local flight on a C-123, an old and loud propeller-driven cargo/transport plane. There was no air-conditioning, and the ride was unpleasantly hot and bumpy. About half of the cadets became physically sick. They told us that once one person "loses it," all it takes is the smell for the rest of the group to succumb shortly afterward. Fortunately, I made it through the flight unscathed. (And I never became airsick during my Air Force career.)

Those of us who had stuck out our program flew to various Air Force bases for a pilot training introductory flight near the end of the summer. My group went to Reese Air Force Base near Lubbock, Texas, where I suited up for a flight with an instructor in a Cessna T-37 Tweety-Bird trainer. It was a short flight, with no aerobatics—take off, fly straight and level, return to the base, and stop.

This was my first flight in a small airplane, and like the A-7 cockpit time, it was eye-opening. The sight of miles and miles of sky, the smell of jet fuel, the feel of the heavy helmet and parachute, and the heat of a Texas summer, all drew me closer to my unusual career choice.

After the flight, I reflected on how we cadets had worried unnecessarily before we flew. *Will we get airsick or not?* The power of groupthink is strong: if you worry enough, you can make yourself airsick just from the stress. In the end, most of us did just fine, and many of us were inspired to make flying our career.

I felt elated when I left basic training that summer. I was proud that I was going to become part of the United States Air Force.

ROTC and Syracuse University

I started ROTC at Syracuse in September 1976. As a junior, I was a member of the Professional Officer Corps (everybody called it the "P-O-C"), which distinguished me and my classmates from the freshmen and sophomores in the General Military Course program.

Our curriculum included a course in national security, taught by Major Mike Lythgoe. He taught us how to scrutinize critically the authors and sources of pieces we were reading. He sometimes broke out in tears during his lectures on patriotism. His profound passion for history and his love of country touched me to the core and earned my respect.

Captain Jim O'Rourke ran our weekly leadership lab. He once told me, "Miss Collins, you are good for at least lieutenant colonel." Perhaps that was his way of saying, "I'm sorry that you weren't selected as ROTC corps commander." I appreciated the compliment, but I wasn't interested in achieving the highest rank or a prestigious position at the time. I was just happy to have an opportunity to serve in the Air Force.

The Vietnam War was still an open wound in the country's psyche. The relationship between universities and the armed forces was strained at best and outright hostile at worst. Although the war had ended in 1975, there was still animosity on our campus toward the military. I loved my uniform and all that it stood for, and I wore it twice a week. It made me a magnet for negative comments and criticism, though. Some students mock-saluted me. Women said, "Ugh! Why are you doing *that*?"

I never let those reactions negatively influence me or change my mind. Those students didn't know what they were missing. I had a chance to serve the country that had given me so much. I also had an exciting future ahead of me: great people to work with, opportunities to fly, travel, good pay, and lots of responsibility. The leadership responsibilities at such a young age were much greater than those available in the civilian world.

I continued following the progress of the women in the Air Force flight training program. In those pre-internet days, my only source of information was the occasional newspaper article. The good news was that all ten women in the first class had graduated and earned their wings. However, I realized that I needed a competitive advantage to get into the program. Gaining actual stick-and-rudder flying experience over the summer seemed like a good option.

The Elmira-Corning Regional Airport offered private flight lessons. I couldn't muster the courage to inquire about them, though. Old doubts ran through my mind. *Will they say no because I'm a woman? Maybe I'm too young.* I could imagine endless excuses for them to reject me.

After weeks of stewing and fretting, I finally mustered the nerve to call the operator at the airport in the summer of 1977. To my surprise, they were very welcoming and invited me to come by. I visited the airport and gladly gave them access to $1,000 that I had saved from my part-time jobs. This would go into my account for flight hours and instructor fees.

"AJ" Davis, a former F-4 pilot who had flown in Vietnam, was the nicest person imaginable and the perfect instructor for me. I finished ground school as quickly as I could, then practiced in flight simulators. AJ introduced me to the Cessna 150, a two-seat propeller plane. He taught me how to perform a walk-around, the preflight "kick the tires and light the fires" check of the airplane. Our first flying sorties that summer took us around the area south of Seneca Lake. I learned to fly turns, climbs, descents, and stalls. We spent most of our time flying around the airport *pattern*—taking off into the wind, turning crosswind, flying downwind parallel to the runway, turning again, making a final approach, and then landing.

After several lessons, I executed a good flight around the pattern and made a nice touchdown. While we were still rolling on the runway, I put my hand on the throttle to add power to take off again. Suddenly, AJ grabbed my hand and pulled the power off.

"I've had enough of this!" he said loudly. "Pull off the runway onto that taxiway over there!"

What? Why is Mr. Nice Guy suddenly so mad at me?

Once we were off the runway, AJ said, "I'm bored, so I'm getting out. You go on—solo."

"Huh?!" was all I could muster.

"You're ready," he said. "I'll just walk in. You'll be fine."

Wow.

I taxied out, feeling excited and grateful he did it this way rather than telling me a day in advance. Otherwise, I could have been awake all night prepping and worrying, worse than for my driver's license test.

I radioed the tower, and they cleared me for takeoff. I applied power, accelerated down the runway, and pulled the plane's nose up. I was in the air, by myself.

Success! My first solo flight!

Just a few feet above the runway, my door suddenly popped open.

My first in-flight emergency!

Focus.

Rather than becoming flustered, I calmly reached over and pulled the door shut. *No problem.* I continued my flight with three uneventful takeoffs and landings. I finished my first solo flight, in my hometown, at the age of twenty. I could scarcely believe how far I'd come in just a few short years.

I didn't have time to complete all the requirements for my private pilot's license that summer. However, having flown solo would definitely give me a leg up in competing for a spot in the Air Force pilot training program.

I attended the ROTC's Third Lieutenant Program that summer with two weeks at Dover Air Force Base in Delaware. They assigned me to the only position available—administrative officer. I felt bored after the first day.

I suggested to another woman in the program, a nurse named Sandy, that we visit the flight line and see if we could get assigned to a flight somewhere. Dover was the home base for the C-5 Galaxy cargo planes, the Air Force's largest planes. We told the clerk at the operations desk that we were third lieutenants and wondered if they had any space for cadets. Jackpot! A flight on Wednesday had space available.

We joined a C-5 crew of three pilots, two navigators, three flight engineers, and three loadmasters for a flight to Andrews Air Force Base in Maryland. A midair malfunction sent us back to Dover. We switched airplanes, flew back to Andrews, and from there onward to Europe. Our airplane required maintenance at every base we landed—Madrid (Spain), Naples (Italy), and Ramstein (West Germany). The planned five-day trip lasted ten days, during which I visited three beautiful European cities, learned firsthand about the military airlift mission, and even had the opportunity to take the controls of a C-5! I returned to Dover just in time to catch a commercial flight back to Elmira.

During my senior year, we toured Patrick Air Force Base and Cape Canaveral, Florida. (I'm eternally grateful to whoever planned those trips for cadets.) We flew the long leg of the trip in a C-130 Hercules, and we unfortunately had to endure airsickness bouts from queasy cadets. It was all worth it once we arrived. We toured the launchpads and hangars, viewed space hardware, and even celebrated with a spontaneous night party on Cocoa Beach.

On the negative side, I missed two days of classes while on the trip. I worried about my grade point average, the primary factor in selection for pilot training. Many other cadets passed up the trip in exchange for potentially better grades. It's sometimes tough to choose between studying to improve your academic scores and participating in an enriching experience. I believe I made the right choice for my situation at the time, and fortunately my grades didn't suffer.

Competing for Pilot Training

In January 1978, I started my final semester at Syracuse. I received orders that after graduation, I would report to Offutt Air Force Base in Nebraska as a computer systems design engineer. As a math major, I would be working in strategic missile targeting.

Colonel Vernon Hagen, our ROTC department chair, called me into his office early in the spring. He rarely spoke to us cadets. I wondered why he wanted to talk with me. He asked if I knew about the test program that the Air Force was conducting to train women pilots, which of course I did. Colonel Hagen then told me that the Air Force was ready to accept women who were recent college graduates into the training program. The Air Force would select up to ten women from ROTC programs nationwide to participate in the next round of pilot training.

He finished with, "Miss Collins, I would like to submit your name for this program. Do you want to do this?"

"Yes, *sir!*" I answered.

"Good, because I've submitted your name already. You will need to complete a physical exam at Hancock Field."

That was undoubtedly the happiest moment of my twenty-one years of life!

That elation didn't last, though. I flunked the physical. My eyesight was not as good as I thought. I was 20/20 in my right eye, but 20/25 for distance in my left eye. The technician disqualified me.

I was heartbroken. I went back to my apartment in tears. I wrote a poem about flight and how much I wanted to be part of the military flying corps. I was unsure about where to go from there.

Colonel Hagen called me back into his office a few days later. "Miss Collins, what is wrong with your eyesight?"

"I don't know, sir."

"Miss Collins, I am sending you back for another test. I want you to rest your eyes, and we'll try again in two weeks."

I bought a bag of carrots. My parents always told me that carrots were good for your eyesight. Over the next two weeks, I ate so many carrots that my fingertips turned orange. (Appropriately, our school's sports teams at the time were called the Syracuse Orangemen.) I cut back on my hours of study and tried to sleep eight or more hours every night.

Two weeks later, I passed the eye exam.

To this day, I wonder if Colonel Hagen told the base to give me a break. If he did, that was not the normal way of doing business. But if anyone ever asks me who most helped me in my career, Colonel Hagen is at the top of my list.

A month before graduation, Colonel Hagen informed me of my selection as one of the eight college graduates nationwide to be part of the Air Force test program for women in military aviation.

What an honor and a challenge, I thought. *Not only do I have the chance to fly for my country, but I can help future generations of women to reach their dreams, as well.*

I wanted to be part of that history. I knew women could do it if given the chance. I never doubted that our group would succeed. This would be just a step toward the future, when even more women could serve the country with their skills and passions.

Most of all, my dream of becoming part of the military pilot community was on the verge of coming true.

I told Colonel Hagen, "I won't disappoint you, sir."

Chapter 3

PILOT TRAINING

After my graduation and commissioning as a second lieutenant, I received my orders to report to the Flight Screening Program at Lackland Air Force Base, Texas, in August 1978.

I crammed everything I owned into my burnt-orange Datsun B210. The drive from Elmira to San Antonio took four days. I arrived during the weekend. The housing office assigned me a room on the third floor of an old barracks.

My class (termed a "flight") comprised fourteen student pilots. We were oddballs, a mixture of aspiring pilots who hadn't previously been through the screening program or who didn't already have an FAA pilot's license. Our flight included four women and ten men, of whom eight were destined for National Guard units and two were Air Force Academy graduates who had missed their flight screening at the academy. We named our flight "The Guard and the Girls." Each class traditionally took a tile out of the ceiling and decorated it. We drew a dollar bill with our class leaders' image inserted in place of George Washington's, in homage to our playing Liar's Poker with dollar bills when we relaxed at the bar.

The other two flights were comprised of Iranian military student pilots from the Shah's regime. Although we trained separately, we interacted with them on the buses, in the hallway, and on the flight line. I wondered what happened to those pilots after the Iranian revolution a little more than a year later.

The Flight Screening Program (abbreviated FSP and nicknamed "Fishpot") evaluated the basic abilities of each potential pilot before sending us to the much more expensive, one-year Undergraduate Pilot Training (UPT) program. The Air Force designed Fishpot to weed out unsuitable pilots, saving the precious UPT spots for those with a strong chance of making it through.

Fishpot tested the basic abilities and attitudes necessary to become an effective military pilot. You need a quick mind. You must be able to simultaneously fly the aircraft, navigate, and talk on the radio while thinking ahead and making adjustments. You need the physical stamina to tolerate turbulence and handle the g-forces of flying. You also need sharp listening skills. It boiled down to a few basic competencies. Do you have sufficient hand-eye coordination to put the aircraft where it needs to be? And can you actively prioritize "aviate, navigate, and communicate" in that order?

Two students in our flight didn't make it through the screening program. One was a woman who had transferred into the Air Force from the Army. She arrived at Lackland with her husband in tow. She dropped out the first day. I was inwardly furious with her, because she took a coveted spot from another woman applicant who also had a dream to fly. But I reminded myself not to be judgmental. You never know what people's personal circumstances are.

Our other dropout was a male pilot who had fantastic potential but became airsick on every flight. I felt sorry for him, because he really wanted to fly. Unfortunately, flying in hundred-degree heat in Texas in August means a bumpy flight. If you have any predisposition toward airsickness, you are done for.

I almost didn't make it through the program, either. It wasn't because of my flying skills. I flunked two medical exams. First, it was my eyesight—again! My left eye was 20/25 for distance vision, but 20/20 was the flight requirement. While I was in the queue, the doctor told me I had a heart murmur. They sent me to the dreaded Wilfred Hall for more tests. Several other medical issues were raised.

The tests were sometimes painful, and my fear and frustration grew with each one.

After a few days of assessments, I either passed the retests or heard that a given problem had been cleared for flight. This was an incredible break. Any one of those issues could have grounded me permanently and changed my life. Who knows where I would be if they told me at age twenty-one that I couldn't fly?

The Fishpot program itself was a great experience. We took the bus every Monday morning to Hondo Airfield, about twenty minutes west of Lackland. We flew the T-41 trainer, the Air Force's version of the popular Cessna 172 propeller-driven plane. We practiced take-offs, landings, basic flight maneuvers, navigation, and communicating via radio. Other than the extremely hot Texas August weather, and searching my plane for rattlesnakes during the preflight checks, I don't remember anything particularly difficult. As flying at Elmira prepared me for Fishpot, I knew Fishpot would hone my skills for the much more challenging year ahead.

I graduated from Fishpot with a *pass*, meaning I had the basic skills necessary to try for my Air Force pilot's wings.

Settling in at Vance

I had requested assignment at Williams AFB near Phoenix for my UPT training. At least two classes of women had already gone through the program at Williams, and it was near a large city. Instead, the personnel office decreed, "You're going to Vance." All I could do was salute and say, "Yes, sir."

Vance AFB, in Enid, Oklahoma, had been training pilots since 1944, but never any women pilots. With Enid being in the middle of the Bible Belt, I anticipated some resistance to women invading a traditionally male profession. And sure enough, I received some interesting feedback on my first trip to the base's commissary (the military version of a grocery store) in early September 1978. I wore my flight suit, and I could tell I was attracting attention as I shopped.

The cashier at the checkout register asked me, in her Oklahoma twang, "Are you one of those new girl pilots?" I told her that I was. She responded, "The wives don't want you here."

This shocked me. I figured some of the men would resist having women in training. However, I didn't expect negative reactions from their wives. I asked the cashier why the women were concerned.

"They don't want you going cross-country with their husbands."

I could understand that. I went home and thought about it. I decided I needed to get to know the wives. I wanted them to be aware that I love flying, I wasn't looking for a boyfriend, and I certainly had no plans to steal anyone's husband!

Administrative tasks consumed my first week at Vance. I filled out paperwork, received my room assignment, registered my car—all the usual activities when you move to a new base. We were issued our flight gear, including checklists, "whiz-wheels" (an aviator's version of a slide rule), and workbooks.

One of the first activities was to have my helmet "poured." Your flight helmet is precisely form-fit to your head to provide maximum protection in case you have to eject from a plane. You don a special hat, and then a hot material is poured into it. When it cools off, the interior of the helmet perfectly cradles your head's contours.

The outside of my helmet was white, and we were allowed to decorate it however we chose. I put a cut-out picture of the bird that was our class logo, and I added $E = MC^2$—Einstein's famous equation and my initials.

The women pilots disliked the fit of the standard flight suit, which pilots call "the green bag." Designed to fit men, the flight suits had broad shoulders and narrow hips. Women needed just the opposite. A suit either fit your hips but was baggy at the shoulders or else fit the shoulders but was skintight on the hips. We asked to have the base's tailor alter the suits. The request was approved—and paid for.

During my first week in Enid, the local TV news stations were covering the new astronaut class of 1978, several of whom were at

Vance for parachute training. This was the first group of astronauts selected in the space shuttle era. Although I never got a chance to see any of them at Vance, it was exciting to know they were there.

Of the six women in the class (Sally Ride, Judy Resnick, Rhea Seddon, Shannon Lucid, Anna Fisher, and Kathy Sullivan), one would someday become the first American woman in space. They were all "mission specialists"—scientists or engineers who would be experimenters, robotic arm operators, or spacewalkers. These women were all role models for me, but none of them would be pilots or commanders.

I began thinking that maybe I could be a space shuttle *pilot* someday.

Preparations

Forty students comprised my Class 79-08—the eighth class scheduled to graduate in fiscal year 1979. The four women in our class (Marlene Brandt, Sue Chlapowski, Lisa Pierce, and myself) were still considered to be part of the test program for women in flight training. Our class also included several Danish students, destined to join their country's air force after they graduated. They were excellent pilots, and also risk-takers. I watched three of them take turns riding a bicycle off the high dive at the officers' club pool. Talk about guts! That was beyond my daring. Our group was split into two sections of twenty, each of which had two women pilots. Most of our male classmates were graduates of the Air Force Academy. Many of them had gotten married after graduation in June and then taken a couple of months off.

One of the guys in our section threw a welcome party at his home on our first weekend. I asked Marlene, the other woman pilot in my section, if she planned to attend. I was apprehensive about going alone, as I didn't know anyone yet. I felt it was important for me to attend, though, especially to meet the wives.

I showed up late. Marlene wasn't there. So much for my moral support! The men were in the dining room, talking about flying, checklists, instructors, and tests. The women were in the living room, discussing curtains in their military housing.

Where do I go first?

The decision was straightforward. I had come here to introduce myself to the wives, so that's what I did. It was probably as awkward for them as it was for me. But I tried to give them a chance to see that I wasn't a threat.

The military is a big family. We needed to support one another. I learned over the years that we don't all have to fit into defined buckets like "male pilot" or "woman spouse." The military culture is now comfortable with women pilots and women leaders, and their spouses have a place to fit in.

However, I did make one rule for myself regarding socialization. I promised myself not to date anyone during the year of training. I didn't want to be distracted by anything, period. I would be "married" to my airplane!

Our first three weeks of instruction included classes on theory of flight, aircraft systems, weather, and pilot physiology. I found the physiology classes to be the most mysterious and intriguing part of the training.

We took turns sitting in an altitude chamber, while a technician pumped air out of the compartment to simulate the loss of pressure in a plane at altitude. I experienced firsthand how depressurization felt.

Another part of the exercise was to drop your mask at an air pressure equivalent to about 18,000 feet altitude. Breathing the thin air makes you hypoxic (low on oxygen). Then you filled out questionnaires while your brain was starved for oxygen. Simple questions like "what year is it?" or "what is your name?" become surprisingly difficult to answer less than a minute after the pressure drops. Seriously—you can forget your own name when you're hypoxic.

This training is critical, because pilots need to be able to understand and recognize the symptoms of hypoxia—light-headedness, dizziness, confusion, gasping—before it's too late to correct the situation. Too many pilots have lost their planes and their lives because they didn't realize they were short on oxygen and slowly passed out.

I'd never been aware of this fascinating information. It piqued my curiosity about how humans can be protected from thin air or outer space, and what it would be like to travel there.

We trained in the use of ejection seats. Our instructors repeatedly told us that the most important aspect of an aircraft ejection was making the decision to eject in the first place. Too many pilots have died due to delayed decision making. Perhaps those pilots feared leaving their comfortable cockpits by being blasted out with the force of fifteen g's. You have to experience how that feels in order to lose your fear of the process.

You practice in a mechanical ejection seat trainer, sitting like you would in your cockpit seat. You then pull two handles to activate the mechanism. It shoots your seat vertically (with you in it) up two long railings. The whole process is over in a matter of seconds. You learn to hold your body in the correct position in order to avoid neck or back injury. Most important, you learn that there is no pain if you do it right.

The other part of ejection training is parasailing, to learn proper parachute landing falls. You wear an opened parachute and a harness, attached to a truck by a lanyard. The truck starts driving, and you begin running. The truck eventually goes fast enough that your parachute inflates and lifts you about one hundred feet off the ground. As the truck drives around the field, you can steer your parasail left or right by pulling on the lanyards. When the instructor signals from the back of the truck, you push two metal bars to release your harness from the lanyard, and you drop back to earth. You try to execute the best possible fall when you hit the ground, by buckling your body and rolling when you land—because if you break your leg, you will not be flying in the near future.

Flight Training Begins

We finally reported to the flight line after three weeks of preparation. I felt, as did every other student, a mixture of anxiety, dread, anticipation, and excitement, since this is where our flying actually started. My section was assigned to "B Flight," which had eight instructor pilots (IPs, pronounced "eye-pease"). The flight commander and assistant flight commander were seasoned pilots with experience outside the training command. The rest of the IPs had three students each. My IP was Captain Rich Murphy. We would be training in the T-37, sarcastically nicknamed the "Tweet" because of the earsplitting screech of its engines.

The IP's job was to make you the best pilot you could possibly be. No wimps, no whining, no complaining, no excuses. You said, "Yes, sir," and did your job, period.

Pilots live by checklists, covering everything from preflight inspection to end-of-flight shutdown. Checklists ensure that you execute the routine steps in the proper order and don't overlook anything.

One of our primary tasks was to "memorize the *boldface*." Checklists contained some steps printed in bold letters. You had to memorize those procedures, because they were so critical that you wouldn't have time to consult the checklist in an emergency situation. What if an engine catches fire? What if both of your engines fail simultaneously? What do you do in the event of explosive cabin depressurization? There would be no time to consult the checklist. You have to immediately execute the memorized boldface procedures.

During our standup meeting each morning, an IP came to the front of the room, described an emergency situation, and called on a random student pilot. "Today the weather is 'clear-in-a-million.'" That meant clear skies and unlimited visibility. "You are solo in Area 2B. You are executing a loop. At the top of the loop, you get a red light on your left engine indicator. Lieutenant Collins, what are you going to do?"

I would stand and start with: "I will maintain aircraft control by rolling to heads-up. What do I see on my engine instruments?"

The IP might continue, "You see high exhaust gas temperature, low oil pressure, and the red engine light."

Given that information, I would recite the boldface procedure for "engine fire in flight."

When you're called on to recite the boldface, you had better get it right. If one word was wrong, the IP grounded you for the day. Messing up your boldface means you are potentially unsafe to fly. You have to write out the procedure by hand, over and over, until the flight commander clears you to fly—hopefully the next day.

We committed to memory the basic biblical words for any pilot facing an emergency or malfunction: "Maintain aircraft control; analyze the situation and take proper action; land as soon as conditions permit." It was always okay to recite that.

I have carried this philosophy throughout the rest of my life to deal with the unexpected. Whenever something unusual, emotional, or unpredictable happens, I go through those steps instinctively. "Maintain aircraft control": maintain control of myself, keep a level head, and calm down the others around me. "Analyze the situation and take proper action": try to figure out what's actually going on— what the real issue is. Don't "chase the gauges." Respond to and correct the cause rather than reacting to the symptoms. "Land as soon as conditions permit": when the situation is under control, get back to a safe and stable place where I can center myself, regroup, and catch my breath.

Focus.

Being a good pilot is all about staying focused on doing your job. It's an interesting combination of complete situational awareness of everything going on around you in this precise moment, while simultaneously thinking about what you are going to do next. If I'm about to start a maneuver, what happens if I can't complete it for some reason? What's the next thing I'll do as soon as I complete the maneuver? If I'm the lead aircraft in a formation flight, I've got to keep my eye on the wingman. What do I do if he suddenly disappears from my view?

You always have to be thinking, *What's next? What's next? What's next?*

When you're flying a high-performance jet supersonically, you can't let your mind drift off onto something that doesn't apply. You can daydream all you want *after* you get home. While you're flying the airplane, you must constantly keep the cross-check going. *Is the aircraft where I want it to be? What's the status of my systems? What's the aircraft doing now relative to the maneuver I'm about to perform? What's next? What's next? What's next?*

As you saw in the story of my early life, I was not born or raised this way. Very few of us are. Fortunately, you can train yourself to achieve this level of focus and discipline.

One fantastic training technique is "chair flying." Grab your aircraft checklist and sit in a quiet room in a chair. Close your eyes. Go through every step of the flight, in order, in real time. Open the checklist and refer to it when you are in the appropriate segment of your imaginary flight.

Start with your walk-around of the aircraft, then go through the engine start procedures. Practice every radio call to the control tower as you request permission to taxi out to the runway and await clearance for takeoff. Picture yourself in every moment of your planned flight profile. *What am I doing at this moment? What's next?* Learn to catch yourself the instant your mind begins to wander. Immediately bring yourself back to the situation and the checklist.

I am certain that chair flying enabled me to perform so well in pilot training. I started with short flights and worked my way up to more complex and longer missions. During each of my imaginary flights, I practiced potential on-the-spot changes to my plan or dealing with in-flight emergencies. When I went out to fly the next day, I felt I was ready for anything.

A whiteboard at the front of the room contained the day's flight schedule, written in grease pencil, with magnetic placards for our names. It was in grease pencil because the schedule changed constantly. You had to be flexible and deal with change. One IP acted as the section's scheduler, usually an IP who was not malleable and didn't care what you thought. You just did what you were told, when you were scheduled to do it.

Before you could fly in a plane, you had to pass the training requirements in the flight simulator, a large box on hydraulic jacks to simulate aircraft motion. It contained a complete re-creation of your plane's cockpit and controls. The T-37 and T-38 cockpit simulators could reproduce six kinds of motion (roll, pitch, yaw; and moving longitudinally, laterally, and vertically) and represented the latest advances in training technology.

The simulators seemed magical. The building that housed them smelled of electronics, and it was cold and dark and somewhat spooky. In the basement of the building were the old Link trainers—outdated but still useful technology. You had to schedule time in the new simulators, but the Link trainers were available to students anytime, without an instructor. After hours, I enjoyed sitting in the Link and "flying" around. They sketched out your ground track on a glass-covered map just outside your window. It was far better than any video game available in 1978!

Instrument training taught me another discipline and mindset that has spilled over into my philosophy of life: the "control and performance concept." It first seemed esoteric and dry when I read it from Air Force Manual 51-37, the instrument flying bible. It reads, "Set the control instruments, cross-check the performance instruments, and then reset the control instruments." In an aircraft, the two instruments that the pilot has any direct control over are the attitude indicator (the *artificial horizon*) and the throttle (engine power). As student pilots, we memorized the attitude and power settings for different phases of flight. For example, your attitude and throttle settings for cruising are different from those when making a final approach to the runway. Once you've put the controls at the proper setting, then you cross-check the performance instruments, which show how the aircraft is responding to the controls—altitude, airspeed, vertical velocity. If the performance instruments don't show what you need, reset the control instruments and watch how the aircraft responds. Don't focus on trying to control the airspeed or altitude directly, or you'll end up in a pilot-induced oscillation, a feedback loop that can quickly make you lose control of the aircraft.

How does this affect everyday life? Since we can't control everything in our lives, decide what you *can* control and set that. You can't make someone do something they don't want to do. Instead, focus on your own actions and attitudes, keeping them intentional toward your goal. See how your actions are affecting the situation and the people involved. Make small adjustments and check the results again. Concentrate on how *you* show up in life.

After several introductory sessions to the T-37 in the simulator, we were ready for the most exciting part—flying the actual airplane. There were three phases to the training: contact (acrobatics, patterns, and landings), instruments (flying with a map stuck under your helmet visor to keep you from seeing outside the window), and formation flying with another aircraft.

I was proud that my performance during training enabled me to be the first person in my class to solo in the T-37, on November 24, 1978. I was therefore the first person to suffer the traditional hazing that followed your first solo flight—the stocks. These were wooden boards with holes that confined your head and wrists. Traditionally, after soloing the first time, the student was placed in the stocks by his classmates, who would take turns wetting this student down with a water hose. While I was thrilled to be the first in my class to solo, I was horrified at the thought of getting my hair wet. I finished my solo flight, got out of the airplane, and went back to my quarters with my helmet, parachute, and other gear. I sat in my room, wondering, *What do I do now?* After about thirty minutes, I knew I needed to take my medicine. It was either now or later. I walked back to the flight room, where the students grabbed me, locked me in the stocks, and hosed me down.

Flying solo in a jet aircraft just five days after my twenty-second birthday has got to be one of the coolest things I've ever done. Where else in the world, other than the military, can a young person have so much responsibility? My flight instructor trusted me to stay safe and to bring this expensive government asset back in good shape. I felt a deep camaraderie with all the other pilots who had done so before me, as I do with all who have come along afterward.

The T-37 phase of the program lasted about five months. I had three checkrides along the way—midphase contact, final contact, and instrument. In those checkrides, an examiner flew in the cockpit with me to assess whether I had mastered the necessary skills. You could fail the check with an *unsatisfactory* by making only one mistake in an otherwise flawless flight. Every pilot feared that.

My midphase check was scheduled for a Monday morning. On Sunday, I drove to Oklahoma City for a shopping trip with the other three women pilots in my class. We ate dinner at a Mexican restaurant. Being from a small town in upstate New York, I'd never eaten Mexican food until I moved to the Southwest. I loved it! Every chance I'd get, I'd order a plate of enchiladas smothered in cheese, with chips and margaritas. I never had any problem with the food.

I woke up on the Monday morning of my very first checkride feeling sicker than I had felt in years. It must have been food poisoning. I had to call in and postpone my exam. I worried that they would think I was chickening out from fear. I kept telling my instructor that I was not faking it; I really was ill.

It was the only time I ever called in sick during the four years I was at Vance.

I learned my lesson. *Be ready to fly. Take care of yourself.*

The next day, I passed my checkride with a score of *good*, which frankly disappointed me. I was hoping for *excellent*.

My check pilot thought I had chickened out the previous day. He told me that I needed a certain mindset for facing checkrides, as every pilot is evaluated routinely throughout their career. He was telling me, "Get used to it."

As I advanced to the next phase, I reminded myself about the control and performance concept: only worry about what you can control! I was happy to score *excellent* on my final contact and instrument checks later.

A Balancing Act

Not everything went smoothly in those first few months. We four women student pilots were ordered to report to Vance several days before class started to give public affairs appearances and perform some other administrative duties. The local news media wanted to interview us, take photos, and get to know us. Colonel Wilson Cooney, our wing commander and the senior officer on the base, decreed that there would be no news media appearances during training. We needed to accomplish all of that before our official start date. Although none of the women cared about the publicity, we followed orders and gave the interviews.

Fast-forward several weeks. Before we even started flying, there was a confrontation in the parachute shop between one of the new women pilots and the squadron executive officer, who was a captain and outranked her by at least four years. He told her, "I just want you to know that *I* never had any special treatment when *I* was a student." He made it sound like she had been asking for the public appearances, when nothing could have been further from the truth. She was so angry at his accusations that she complained to our class commander. The executive officer lost his position and was demoted back to a basic instructor pilot.

The whole situation seemed very unfortunate. Base leadership was trying to send a signal to the men not to harass the women. Instead, it created the impression that our presence was an intrusion.

I resolved to try to handle any future incidents myself, person to person. I would only escalate them if they became a recurring issue.

Fortunately, I never had to deal with clearly improper behavior from any of my colleagues.

The presence of women pilots had a positive effect on morale. Every once in a while, a male pilot might even admit, "I'm glad the women are here. People are more courteous. Their language has cleaned up. Even the guys are trying harder."

Sometimes people would kid around in a good-natured way that showed that we women pilots were still a novelty. One day, I was practicing formation flying as a wingman in the T-37, and the lead pilot and I made a quick call to each over the radio. Air Traffic Control called back, "Curly 2-1, is your wingman a wing*woman*, or is his seatbelt just too tight?"

I didn't want to change the culture. I wanted to fit in. I did as much as I could by getting along with others, helping out when I could, talking things through with my classmates when they had a tough time in training, and hanging out at the officers' club on Friday nights.

Unwinding after an intense week of hard work was a particularly enjoyable part of our training. We'd get together to socialize, tell "there I was" stories, and dance. If a tornado warning forced us out of the squadron, we'd go over to the O Club, grab a beer, and watch the clouds swirl and turn green with hail.

Training in the Talon

We began the T-38 phase of our training in March 1979. I found the T-38 Talon to be far more intimidating than the T-37. First of all, the T-38 was supersonic. The T-37 maxed out at about 425 mph; the T-38 could exceed 850 mph, or 1.3 times the speed of sound. It was longer and weighed nearly two tons more than the T-37. Instead of the T-37's side-by-side seating, the T-38 pilots sat *tandem*, with the student in the front seat and the instructor in the rear.

The first time I took off in a T-38 with my instructor pilot, First Lieutenant Bob McHale, I thought I had left my brain back on the runway.

I wondered, *How will I ever get used to flying this white rocket?*

I eventually learned to love the airplane—knowing that it was also the airplane NASA astronauts used in training.

As with the T-37, the T-38 program included contact, instrument, and formation flying phases. The initial goal was to get students "soloed out," so we practiced touch-and-go landings until our instructor felt comfortable letting us out alone with the plane.

Keeping track of up to twelve aircraft in the flight pattern was incredibly difficult. To avoid midair collisions, we had an air traffic controller in the tower and an instructor in the runway supervisory unit (the RSU, also called "the box") out near the touchdown zone. The sharpest instructors sat in the box and acted as controllers. A quick call of "Flare! Go around!" would warn a student to abort a landing approach if things looked dangerous. Each student had some help, but basically when conditions were clear enough for visual flight rules (VFR), each pilot was responsible for maintaining separation from other planes.

We flew out west of the base to designated "flying areas." These were zones where we were given the airspace between 10,000 and 23,000 feet in altitude, because it took 10,000 feet to do a loop in the T-38. We flew aerobatic maneuvers including loops, cloverleafs, Immelmanns, lazy eights, split S's, barrel rolls, and aileron rolls.

When we entered the formation flying part of the training, we learned to keep stable "on the wing" of another plane with only three feet of separation between our wingtips. We practiced two-ship and four-ship tactical formations, learning how to provide flight support to the lead aircraft flying the main mission (such as a bombing run).

Formation flying is the ultimate embodiment of energy management and situational awareness. For example, a basic formation maneuver is the "rejoin." The lead aircraft would fly a stable thirty-degree bank at three hundred knots. The wingman would separate off to some distance then try to rejoin on the wing of the lead plane. If the wingman had too much speed, too much altitude separation, or too much difference in heading, the planes could collide. Both pilots learned how to carefully manage their plane's energy and position. In the event of a misjudgment, the "overshoot maneuver" had the wingman give up the rejoin and move to the other side of the lead plane. You needed to know how to execute the rejoin quickly and efficiently, especially if you were approaching a cloud bank and might lose sight of the leader. Patience and discipline were the key.

The instrument phase primarily involved flying approaches to the airfield "under the bag." To practice for flying in clouds or adverse weather conditions, we had a "hood" that we could pull up on bungee cords inside the canopy to block our outside view. The IP, sitting in the other cockpit, could look out his window while the student flew using only instruments.

We flew to several bases in Oklahoma and Texas during our training. We relied on paper maps and "approach plates" (maps with procedures for making an instrument approach to an airfield). GPS hadn't been invented yet, and back in the old days, pilots had to shuffle papers scattered all over the cockpit, while holding altitude, steering in the right direction, and talking on the radio. Nowadays, an electronic screen displays everything for the pilot, and human factors engineering—improving the layout of the cockpit and reducing the potential for distracting or confusing information—has made flying much safer and more efficient.

Toward the end of our yearlong training, the student pilots began anxiously awaiting their operational flight assignments. The instructors had been meeting over the months to review each student's capabilities. Instructors assigned some of us a "FAR" rating (for fighter/attack/reconnaissance), which indicated that they felt we had sufficient knowledge and skills to become an aircraft commander. Other students received "TTB" (tanker/transport/bomber) assignments as copilots. I received a FAR rating, which qualified me to be an instructor.

The "assignments night" event at the officers' club came about four weeks before graduation. Instead of learning about your assignment from your flight commander in his office, the announcements were made in public with your spouse and classmates and every other student on base in attendance. The room was packed, and the beer flowed. Each student was called upon one at a time, stood up, and a photo of their assigned aircraft was projected on a screen.

It was a big event—the start of your career as an operational Air Force pilot.

I had a hunch I might be staying on at Vance as an instructor pilot. My flight commander, Major Slayton, had called me into his office a few weeks earlier. He told me why I would not be a T-37 instructor. It seemed a rather silly reason—something about them not being ready for a woman in that section yet. "But we here in T-38s are ready!" he added. He told me to keep the conversation between us.

When they called my name on assignments night and a photo of a T-38 went up on the screen, I was not entirely surprised.

I walked home that evening with mixed feelings. I was happy to continue flying the White Rocket. It also began to sink in that my training—a part of my life that I would cherish forever—was coming to an end. Memories flashed through my mind: remembrances of successes and failures, happiness and stress, organization and scheduling chaos, working tremendously hard and seeing progress every day.

I had made it this far. After just a few more flights, I would be one of the first women to receive her flying wings at Vance Air Force Base.

Chapter 4

STRENGTH AND HUMILITY:

TRAINING PILOTS

Before I could become a full-fledged operational pilot, I had to complete several other vital and challenging programs.

The first was the weeks-long Survival, Evasion, Resistance, and Escape Program at Fairchild AFB, near Spokane, Washington. In the event of a crash landing or ejection over the wilderness, a crew member had to know how to stay alive—and motivate the rest of her crew to survive, as well. We started with several days of classroom training, taught by top-notch enlisted airmen to whom you would willingly entrust your life.

Instructors separated us into groups of eight people, and we set out on the trek. Two instructors accompanied us through the wilderness, in a scenario that realistically simulated the aftermath of an airplane crash. We were each issued a forty-pound backpack of equipment, a very limited amount of food, and only the clothes on our backs. We would have to find water along our way, either in creeks or else by building a solar still to condense water from the humidity in the air. We hiked many miles each day through the rugged terrain of eastern Washington state.

I was the only woman in our group, and that was okay with me. Our team included officers and enlisted personnel—fighter pilots, cargo pilots, weapons systems officers, flight engineers, and loadmasters. An older loadmaster in our group was a smoker and in poor physical condition. The rest of us took turns carrying the forty-pound pack that he couldn't manage himself. That meant that one of us—including me—was carrying eighty pounds of gear at one time or another.

I don't think I appreciated the mental conditioning part of the challenge until we were on the trek. Yes, we learned how to eat, sleep, breathe, and treat injuries that we might face, but we also had to learn and experience the psychological problems we would encounter. They taught us to fight our inclination to ration water; we were actually supposed to drink it early to avoid dehydration. We tried to fight our natural revulsion to certain foods. The instructors caught, killed, and skinned a rabbit right before our eyes—and then tried to give me the intestines to eat.

Each student had to prove that they could eat some sort of unpleasant food, or else be hounded until they did. I figured this out right away. I searched for something that would prove my worthiness, so that they would leave me alone when it came time to eat the nastiest stuff. We came upon a dead log filled with termites. "Watch this," I said. I picked off two termites and ate them as everyone looked at me. *That should keep them off my back*, I thought, and it worked. Nonetheless, we seemed to be hungry all the time.

Our instructors taught us about the importance of having the *will* to survive. If you sincerely believe in your abilities and potential, you can overcome overwhelming odds. We learned about actual incidents where one person survived while others perished, despite being in the same situation with the same resources. How could this happen? Determination can drive you above and beyond, if you possess the will to succeed and the will to survive. We were challenged to think and act clearly, despite feelings of heat, cold, hunger, or fatigue.

I stayed motivated. I did not want to be "that woman who failed." I also loved the wilderness and felt part of it.

Thank God for Camp El-Ne-Ho!

I reminded myself that for centuries humans have survived off the land, struggling through famines, droughts, plagues, infections, insect infestations, stranger danger, weather disruptions, and more. *Surely, I can survive a few weeks of this. So what if I can't shower or cut my finger-nails for weeks? So what if there is a scorpion in my sleeping bag? So what if I can't sleep because my feet are cold and my teeth are chattering? This is an opportunity to learn more about myself as a person—to find out if I am tougher than I think.*

I rationed my food until late in the training. The other officer in our small group, Bob Vosburgh, caught and collected the grass-hoppers that seemed to be everywhere. One evening he boiled a pot of water, tossed in the piece of beef jerky he had been saving, and then put the mesh bag of grasshoppers into the water. He and I ate them, much to the disgust of the others in our group. That protein boost made a huge difference for us the next day. I saved my can of tuna for the last day of the trek, knowing that I would need all my energy for what was coming next: prisoner of war (POW) camp.

Captured!

At the end of the trek, we were all captured by "the enemy." They tied our hands behind our backs, put burlap bags over our heads, and marched us to a POW camp. I was surprised at how alert my hearing became when I couldn't see. Some light came through a small rip in the bag over my head, and I was able to nudge the bag enough with my shoulders to peek through the rip. I saw that our captors were actually our instructors, dressed as Cold War–era Eastern European guards.

Every one of us received harsh treatment. They pushed us around, cursed us, called us names, and treated us roughly. They assigned each of us a number. They never used our names. I was simply "Number 36." I'll remember that forever, because they used it so often to slander me!

There was one other woman in the group of perhaps one hundred of us. We never talked; I was only aware that she was there. We were both constantly called "wench." They clearly meant it to offend us, but I had never heard the term.

The guards shoved us into individual small cells, about six feet square, still with the bags over our heads, and left us by ourselves for hours. This was the "self-isolation" part of our evaluation. My cell had a bucket, for obvious reasons. From what I could see through the rip in my head covering, the rooms were in a hallway set up like a jail. The instructors made terrifying noises, like they were torturing people. I never actually *believed* that they were torturing my fellow prisoners—that would not have been legal—but it sounded real enough to heighten my natural fears.

At one point a guard yanked me out of my cell. He noticed that I could see him through the rip in my head bag. "Filthy wench!" he yelled at me, and then he replaced it with a new, much darker bag. *Darn!*

Before we embarked on the trek, the instructors had told us that anytime a student felt they were in a situation that seriously and truly threatened their safety, the student should call out the safe words, "flight surgeon." Calling out that phrase stopped the scenario, and safety became the only focus. During the first day of our POW experience, several of us were yanked out of our cells—bags still over our heads—and pushed into tiny boxes. My box was so small that I had to squish myself into a fetal position. I don't know how long they left me in the box—maybe several hours. I assume they wanted to see if I was claustrophobic. I was actually quite content in there, as I could nap and chill out. There was airflow, so I wasn't in physical danger.

The loadmaster from my Trek group—the smoker who couldn't carry his pack—was shoved into a box next to mine. After a minute, he began to sob and yell, "Help! Help! Get me out of here!" I whispered to him, "Are you okay?" He ignored me and kept shouting. The guards came by and ordered him to be quiet. He continued to cry out.

A few minutes later, I whispered to him, "Do you need the flight surgeon?" That got his attention. He yelled, "Flight surgeon! Flight surgeon!" The guards removed him from the box.

Several hours later, he was back in the box again. He immediately yelled, "Flight surgeon! Flight surgeon!" I felt bad for him, as he was obviously severely claustrophobic. The guards pulled him out. I never saw or heard from him again. He did not pass the training course.

After a day of this mistreatment, they released us into the open compound. This scenario was patterned on well-established POW camps, where, after initial capture and harassment, guards would put POWs to work. They removed our head coverings and formed us into groups to perform meaningless manual labor.

In our initial classroom training before the trek, our instructors had told us that we would be in this camp scenario and that we should try to escape. In fact, they threatened to extend the torturous program if we didn't at least *try* to get out. The rule in this scenario was that only one person needed to escape the camp, but the planning and execution should be a group effort. The person making the escape had to carry a special piece of paper (the "chit") with him when he escaped. The chit included a map or directions where the escapee should go once outside the fences. The destination would look like an old-fashioned phone booth. Once inside the phone booth, the escapee would call the phone number on the chit, and the group would be declared as having "won" with a successful escape.

While performing our menial tasks in the yard, someone in the group got hold of the chit. We decided that some of us would start a fight to try to distract the guards, and then the guy with the chit would climb over the chain-link fence and run away. That was our elaborate plan.

Meanwhile, we grew weak from hunger and fatigue. No one had eaten a real meal in at least two days. Nobody had slept, because the guards woke you up every time you nodded off. We were in extreme mental and physical distress, just like we would be if we had been taken prisoner.

This was the crucible, where we'd really learn about ourselves. How would we react? Would we turn into vegetables and mindlessly follow every order? Would we fight back? Would we pass out? Would we risk our lives to try to get out? The fatigue was so overpowering that you almost forgot that it was a training scenario.

Hours went by. After dark, with no end of our mistreatment in sight, the guards opened the camp gate and left it open. *Okay,* I thought, *they are encouraging us to get around to it, to do our thing and try to escape.* Others took notice, too, and finally the appointed POWs started their fake fight. Because we were so exhausted, it was the weakest excuse for a fight I'd ever seen! They were just going through the motions. I was happy that at least something was finally happening.

The guy with the chit just stood there and watched.

I couldn't believe it. *Why is he not running out the gate? It's wide open!* He just stood there, doing nothing, his eyes glazed. I could see the chit in his hand. He must have been so drained that he had forgotten what to do.

I ran over, grabbed the chit out of his hand, and dashed out the open gate.

I sprinted as fast as I could across the field. A siren wailed. I continued to run. A searchlight swept across the field. I kept my eyes on it as I ran. When it came close to me, I dove into the grass and lay motionless. It moved over me and kept going. I got up and ran again as fast as I could. The searchlight came back again, and I dove into the grass. This time, the light stayed on my back for several seconds before passing on.

I ran into the pitch-black woods. I had no flashlight, and I could see nothing because my eyes hadn't adjusted to the darkness. I walked quietly along a rock wall. *I need to get somewhere so I can read this chit,* I thought. *I have no idea where I'm going.*

Just then, I was jabbed in the back with the butt of a rifle. "Put your hands over your head, filthy wench!"

I failed.

The guard marched me back across the field. It seemed to take forever. We entered the gates and the bright lights of the camp. I still had my hands over my head, and the other prisoners whistled the theme song from the old movie *The Great Escape*. The guards tied me to a tree, where I stayed until sunrise.

They marched me into the office of the camp commander. He came in and had his guards untie my wrists. He flirted with me and tried to get me to betray my comrades. He promised me food and water and a nice place in the hierarchy of the POW camp if I defected.

I just sat there, thinking how hungry I was and how disappointed I was to have failed the escape. Since I wouldn't acquiesce to his demands, guards tied me to the tree again, where I stayed for the remainder of the scenario.

As the day started to warm up a bit, I wondered if someone would attempt another escape. The camp commandant eventually came out and ordered all the prisoners to line up. I was untied from my tree, and I joined my fellow captives. The camp commandant started chewing us out, telling us what a group of losers we were.

Then he switched his tune.

Suddenly, he was talking about the United States of America being a magnificent country, where people are independent and free.

What is this? I wondered.

Then he yelled, "About face!"

We all turned around, heard a loud *boom!*, and were surprised by fireworks. Right there, in front of us, was the most beautiful sight I had ever seen: a tall, huge, bright garrison flag! An American flag! With bright lights shining all around us. The national anthem started playing over the loudspeakers. We all saluted our flag, and I don't think there was a dry eye in our group. This, officially, was the end of our training!

I learned so much about myself from this experience. Although the training varies and has evolved over the years, I believe all attendees come away from military survival training with common experiences and wisdom. We learn how to survive in the outdoors and hazardous conditions, with no comforts of life. Whether in the desert, Arctic, forest, jungle, or mountains, we gain confidence that we can live off the land, safely and for a long time. We learn to resist the inclination of self-defeatism. We learn the *will* to survive.

In the case of the POW camp, we learn to let go of things we don't have control over, to separate what we can and can't control. Those who emerge from the program as leaders are the ones who have the necessary knowledge and the positive attitude to keep the group cohesive and alert. A smile, an encouraging comment, or even some light humor can go a long way in achieving success. We also learn to stay true to our fellow crew members and our country. And we learn that no matter how bad things get, the situation can turn around quickly.

Be prepared for anything.

Training to Train

Immediately after completing the land survival course, I entered a four-week program, back at my home base of Vance, designed to transition a new T-38 instructor pilot to the skills needed to fly the plane from the back seat. Unlike the introductory T-37, which has side-by-side seating, the tandem T-38 has the student in the front and the instructor in the back. You can't communicate with body language. You must rely on verbal communication—assuming your intercom is working. The new instructor focuses on learning to land with different configurations of flaps up or down, engine out, and crosswind landings. Visibility is limited from the back, but with enough practice the instructor is soon an expert at showing students the "right picture." This short program prepared us for the full-blown Pilot Instructor Training (PIT) program at Randolph AFB in San Antonio, Texas.

I arrived at Randolph on November 4, 1979, the same day the US embassy workers were taken hostage in Tehran. I watched the situation in Iran with personal interest, since I had met the Iranian pilots who trained with us in 1978. Every night's newscast started off with the number of days the embassy personnel had been held hostage. I routinely kept up to date on national and international defense and diplomatic issues, because I knew that world events would affect my military assignments. The situation in Iran felt even more personal to me, though.

I had to put those events out of my mind as best I could, because my immediate mission was to graduate from the PIT course. This four-month program makes a pilot proficient in the aircraft from both seats, and it provides experience flying increasingly challenging sorties.

I don't recall learning anything about how to be a good teacher. We concentrated entirely on our flying skills. The philosophy could be summed up as "a picture is worth a thousand words": show the student what *good* looks like. We were assessed on our ability to talk and fly at the same time. I never had any trouble talking while flying; my challenge was actually to talk *less*. Most student pilots can't concentrate on flying while listening to their instructor, so being a motormouth instructor can be a liability. The best instructors select a few choice words and bark them out over the intercom to their students at the perfect moment.

San Antonio's winter weather wasn't particularly cold, but we frequently suffered low cloud ceilings that created difficult navigation conditions. We relied on radio signals and course and bearing pointers to determine our location over the ground. The challenge was to stay within your designated flying area without drifting into an airspace violation.

While flying in formation, the wingman had to maintain three feet of wingtip-to-wingtip separation from the lead plane. This got particularly exciting as the two planes moved in and out of clouds, giving you the sense of tremendous speed as well as unexpected vertigo. To a pilot, vertigo is losing your sense of which way is up. When you're in the clouds and focused intently on staying on the wing of the lead plane, you can feel like you're in a turn or even flying upside down, while you're actually flying straight and level. You must learn to disregard your feelings and follow your eyes. Believe what you see, not what you feel.

Toward the end of PIT, I felt comfortable in the airplane. It was as if I was strapping the airplane onto *myself*, not strapping myself into the plane. The T-38 became an extension of *me*. The control stick, throttles, switches, and buttons were just ways to make the airplane do exactly what I commanded. The more I flew, the smaller the airplane felt, and the more confident I became.

The T-38 could perform just about any maneuver a human body could withstand. "Over the top" maneuvers required a 500-knot starting speed, followed by a 5-g pull on the stick. Flying that maneuver stressed your spine and neck, and we would sometimes execute it up to ten times a day. The airplane itself was limited to about 7 g's. If you flipped upside down and pushed on the stick, you'd generate negative 1 g. In that case, all your blood would rush to your head. The aircraft couldn't withstand negative g's for long. The engines would fail because the oil pumps would cavitate, creating bubbles that wreak havoc on the pump mechanism. The T-38 could execute an aileron roll at 720 degrees per second—two complete rolls around the plane's long axis in one second—which quickly disorients the pilot.

On top of that, we practiced many of these maneuvers with a wingman just three feet away. Flying as lead pilot required me to be smooth and deliberate, planning ahead to ensure that I had the required ten thousand feet above me and that I'd still be going at least two hundred knots at the top of the loop.

My favorite maneuver was the cloverleaf, four loops that were at right angles to one another. I had to manage the T-38's energy so that I hit 450 knots at the bottom of the loop and 175 knots at the top, while making accurate turns to line up 90 degrees after each leaf.

My next class was water survival training at Homestead AFB, south of Miami. This was a relatively short program to teach a pilot how to survive ejecting from an aircraft and landing in a body of water. We learned how to steer a parachute, execute a parachute landing fall in the water, inflate a life raft, bail out water, repair rips and tears in the raft, and signal our rescue forces. We had to repeat the process over and over until it became a habit. As with memorizing the boldface in our checklists, there's no time to try to figure it out in the event of an actual emergency. As a fun capstone to the water exercises, a rescue helicopter hoisted us up from our rafts in the ocean.

I wanted to make the most of my brief visit to Miami. I had booked myself on the latest possible commercial flight home, hoping to rent a car and drive to Key West. It was a much longer drive than I realized. I only made it about halfway there and then had to turn around and head back.

Then came a lesson in comeuppance. I had a flat tire, and I missed my flight. I had to spend the night in a hotel and then fly back the next day to my new instructor job at Vance. I hustled into the squadron building that afternoon and reported in. My flight commander calmly told me that his boss, the squadron commander, wanted to see me—never a good sign. He chewed me out for returning late, reminding me that there was important work to do, and so on. He was absolutely right.

My first real job in the Air Force, and I was getting off on the wrong foot.

Lessons in Humility

In March 1980, I began my three years of flying T-38s and teaching young, aspiring pilots how to be the best they can be. I was excited and ready to go! I was assigned to "O Flight," and Major Gary Thompson was my flight commander. He was a great guy, a calm and wise leader. He never raised his voice or overreacted. His job was to oversee eight instructors and more than twenty students.

My role was basically sitting on the other side of the table from where I had been for the past year. Now I would be teaching young pilots the maneuvers and philosophies that I had just finished learning myself. My unofficial title was first assignment instructor pilot, or FAIP (pronounced "fape"). About half of the instructors were FAIPs, and the rest had previously flown operationally. The operational pilots were there to show us what it was like in the *real* Air Force.

In my second year, the base's new wing commander, Colonel Giles Harlow—a former pilot of supersonic reconnaissance RF-4 aircraft—called me into his office, and I guessed he must have been curious about his only female flight instructor. I saluted in a military manner, wondering why he had wanted to see me.

"Sir, Lieutenant Collins reporting as ordered."

"Collins, how would you like to be the first woman to fly in an F-15?"

I was astounded. "Huh?" was about all I could manage to say.

"I have you set up for a trip to Holloman AFB for an F-15 flight."

This should have been exciting news for me, but it was not. I knew of the exchange program for senior FAIPs to attend one day of instruction and a flight in the latest aircraft as a reward for three years of hard work as an instructor. Only the sharpest FAIPs were considered for this flight, and usually only one pilot per year.

"Sir, isn't that program for senior pilots who have paid their dues?" I asked.

"Do you want the flight or not?" he barked.

"Well, yes sir, I do."

"Fine. It's all set up."

I went back to the flight room and told the guys what had happened and how I felt about it. Most of them seemed understanding and supported me. However, I could tell that many were cynical about a woman getting an unfair break.

On December 4, 1980, the F-15 landed at Vance to pick me up for my day at Holloman AFB in New Mexico. The F-15 was a new and advanced airplane in 1980, and students rarely saw one. It attracted a lot of attention at Vance.

After a briefing with Captain Yates, the cool and sharp pilot, we taxied out to Runway 17 and were cleared for takeoff. I hit the afterburner—all five stages of afterburner. I think I had counted to two when I was already off the ground, gear up, pulling 5 g's. I flew nearly vertically to 20,000 feet, rolled 180°, and leveled off to the horizon at 24,000 feet. I was still over the base. I looked down and saw the runway directly below me. The F-15 felt like the closest thing to a rocket that existed.

Incidentally, not long after my flight, Stage 5 afterburner takeoffs were disallowed. There was operationally no need to use such tremendous power.

We arrived at Holloman's airspace and engaged in mock air-to-air dogfighting, which Captain Yates flew. Then I observed a debriefing that consisted mostly of swearing and cursing. Frankly, I'm not sure I followed any of it.

Before I left Holloman, Yates asked if I had any questions. I told him about how I was selected for this coveted flight, over the more experienced FAIPs. He responded, "You're *not* the first woman to fly in an F-15. We've flown women photographers before."

So much for making history.

However, although I can't prove it, I'm pretty sure I was the first woman *pilot* to actually fly an F-15.

—

Colonel Harlow was an interesting person. He certainly knew what he wanted to do, and he never hesitated. Our local call sign for T-38s was "Reno," and our cross-country call sign was "Duke." I loved the Duke call sign—tough and straightforward, just like John Wayne himself. A few years into my assignment, Colonel Harlow decided to change this venerable call sign to "Vandy." I have no idea why he did it, other than that he *could* do it. Vandy sounded weak and frail, in my opinion.

I thought I would test the system during a cross-county flight to Langley AFB, where I had a friend who flew F-15s. I told my friend about the wimpy call sign. He and I used to discuss the old Soupy Sales TV show from the 1950s and 1960s, which had a lion puppet named Pookie. I decided to change my Vandy call sign to Pookie for my trip home. This was not normal procedure, but as far as I knew, no one said we couldn't do it. I filed a flight plan to Grissom AFB in Indiana using the Pookie call sign, and once there I would change the call sign back to Vandy for the trip back to Vance. I thought no one would find out, since this was a Sunday afternoon.

It was fun for about two hours.

When we landed at Grissom, base ops told me that the supervisor of flying from Vance was looking for me. Vance had been asking for Vandy 68. The Grissom ops guy reported there was no Vandy 68, just a Pookie 68.

Long story short, the Vance ops officer called me on the carpet Monday morning and chewed me out. He told me that I might have accidentally used a classified call sign.

"OK, yes sir," I said.

As I walked out of the office, I could see him and his executive officer laughing. So much for discipline.

That incident may seem harmless, and in fact it was. However, I had to relearn humility many times throughout my three years as an instructor. Humility can keep you from getting killed.

I my first month as a FAIP, I was assigned to fly with the top student in the section. I figured out that this guy was so good all I needed to do was to sit in the back seat and watch. Heck, I could probably just sit on my hands for the entire flight!

We went up in our T-38 to practice formation flying. In the middle of the flight, while we were the lead aircraft, our wingman began executing a straight-ahead rejoin, coming up on our left wing. At that moment, my student said to me, "The edge of the flying area is close."

I responded, "Do what you need to do to stay in the area," thinking that he would start a gradual turn.

Instead, he said, "Ninety left, *now!*" He rolled fast and made a hard turn to the left—directly into the approaching wingman.

Fortunately, the instructor in the wing aircraft saw the maneuver and reacted instantly, pushing hard down on his plane's controls. We narrowly avoided a midair collision. It was probably the closest I ever came to dying in an airplane. My first month on the job.

That was the first time my student received a failing mark in pilot training.

I felt as if I could have prevented the incident if I had been more alert. We all learned a tough lesson: *Never, ever let your guard down.* Even the sharpest and most respected people can make mistakes.

My humility was restored . . . for a little while.

It was easy for a FAIP to get a big head. It seemed like we were always saving students from killing themselves. In that circumstance, a FAIP can develop the attitude that she is the best pilot who ever walked on planet Earth.

Despite my early lesson in humility from my first month on the job, I let myself get sucked back into an attitude of superiority and infallibility when I received an early and rare promotion to Check Section. Check Section is an office of six pilots whose job it is to evaluate students on checkrides at various milestones in their training.

At that time, I was the youngest pilot assigned to Check Section at Vance. Unfortunately, I became overly proud of myself.

The flying profession has a way of knocking a big ego back into place.

I was on a cross-country flight once with a student, returning to Vance from Memphis, Tennessee. There were thunderstorms along the route. Planes didn't have Doppler radar back in those days. Instead, we used preflight charts with the radar returns drawn in hashed lines to show where the thunderstorm cells were located. Those charts were usually hours old, and Midwest weather can change dramatically and quickly.

Before we took off, my student and I agreed we should be able to see far enough ahead to fly around the storms, and all the other instructors also decided to fly home that afternoon. Unfortunately, our T-38 was one of the last planes leaving Memphis. As we flew west at 33,000 feet altitude, we saw a storm ahead of us. We called Fort Worth air traffic control.

"Fort Worth Center, Vandy 33 requests flight level 370." (That would be 37,000 feet altitude.)

They denied our request.

"Fort Worth Center, Vandy 33 requests deviation to the right around this thunderstorm cell."

They also denied that request.

They denied every deviation I requested from our flight plan. My mistake was that I failed to declare an emergency. If you declare an emergency, they must approve anything you ask for.

Instead, we flew into a hailstorm.

Lightning flashed all around us. The sound was deafening, like guns blazing and ice cubes being hurled at our acrylic canopy. The turbulence was so bad that I took control of the plane from my student, and I could not hold altitude plus or minus two thousand feet.

It was the most frightened I have ever been in an airplane.

Dear God, please get me out of this, and I promise I will never let it happen again!

We eventually popped out of the clouds on the other side. Neither our altitude indicator nor our vertical velocity indicator functioned. Fortunately, we still had a working airspeed indicator.

I finally declared an emergency, requesting a deviation to Sheppard AFB near Wichita Falls, Texas.

We landed safely and taxied to the apron. Maintenance personnel awaited us. The airman climbed up the ladder to the cockpit and gave me a glass of water.

"Drink this, ma'am. You're gonna need it."

Uh-oh, I thought.

I climbed down the ladder. I was just happy to be back on solid ground again. Then I looked at the airplane. The leading edges of the wing were dented. Hail had punched a hole through the plane's nose cone. The metal pitot tube, which measures altitude and airspeed, was chipped and damaged. That's why several of our instruments weren't working.

I had trashed this beautiful plane. To add insult to injury, after I called Vance and reported the condition of my airplane, the squadron had to send two planes to bring us home the next day.

I had to stand up in front of the entire squadron—instructors and students alike—and explain how I had gotten myself into this situation. It was standard practice to have a pilot debrief any kind of unusual incident or screwup with the squadron so that others would not repeat the mistakes.

I firmly stated that I learned to be aggressive in declaring emergencies. Don't let air traffic control dictate your safety. The pilot can best assess the local situation and needs to be in charge.

Aside from my bruised ego, my student and I were uninjured. The airplane was repaired. My career advancement continued on course.

The squadron commander shared some final words of wisdom: "Any pilot with a star on his wings has been in a thunderstorm. It's what you learn from it and what you do with that experience that matters."

What the Instructor Learned

Here are some of the other lessons I learned as a flight instructor. Many are applicable to everyday life, too, as with a parent teaching a teenager to drive the family car.

Go to the aircraft hangars and talk to the maintenance guys. After reading the manual one hundred times, any serious pilot still wants to learn more about the airplane. I visited the hangars to look at planes that had panels removed, so I could actually see the electrical lines, hydraulic lines, generators, engine blades, avionics boxes, and all the other systems. Talking to the maintenance people helped us both do a better job.

Don't talk too much in the airplane as an instructor, or your student will stop listening.

Give your student increasing responsibility. Have them plan their own flight, brief you on their plan, and include options in case the plan changes.

Ask your student thoughtful questions and encourage them to talk through potential emergency situations.

Never, ever let your guard down. Complacency can kill.

Never, ever trust the weather forecast. Always have a backup plan.

Sometimes, the best instructors are the ones who have struggled themselves, as they can anticipate where the student has a misunderstanding.

Don't take things personally. Have a sense of humor. Our first wing commander liked to start his "Commander's Calls" with "Gentlemen . . . [pause] . . . and lady," talking about me, of course. I could have let it upset me, but why? I just laughed and let it pass.

The instructor's attitude rubs off on the student. This is especially true for parents and their children.

Follow the rules. I felt obligated to do so, even though many of my colleagues ignored them. I did not want to hear, "The woman broke the rules and screwed up."

I was the only woman instructor in Vance's T-38 squadron from September 1979 to December 1982. The T-37 squadron finally brought aboard a female instructor, and eventually three women instructors worked that side of the program while I was stationed there. They all did an excellent job, and I am sure their students received top-notch instruction. One month after I left Vance, another woman was selected to train T-38 pilots. There have been many more since then.

Women have proven themselves as excellent pilots as well as instructors, and their male students have become excellent pilots while working with them.

I left Vance after slightly more than four years. Although I had not asked for Vance—and had actively tried to change my initial assignment there—I am so glad I had that experience. I learned to love Oklahoma. The air was always fresh, and the weather was powerful. Pilots develop their best skills by dealing with extremes in weather, like strong and shifting winds, snowstorms, low clouds, and extremes of temperature. The prevailing winds were so strong that the few trees all leaned to the north. The unmarried pilots joked, "There's a girl behind every tree in Enid—but there are no trees!"

Oklahoma exudes aviation. I made friends with Bob Eadie, another instructor, who let me fly his Cessna 140, a vintage single-engine "tail-dragger" plane from about 1942. He kept his Cessna at Woodring Airport, near Enid. Its central location made it convenient to fly to air shows and competitions around the state on weekends. I enjoyed the competition and camaraderie with the pilots I met along the way. Bob checked me out on his motorcycle. It was great fun to run the wide-open Oklahoma roads and trails while exploring the state's natural beauty.

Most of all, I loved the people of Oklahoma. They seemed fearless, open, and welcoming. The people of Enid clearly cherished having Vance in their town. There was a close and special relationship between the civilians and the military personnel, even though the individual personnel rotated out every three or four years. I met and became friends with many of the local folks in the base office buildings, my church, the social center, and the officers' club. They supported us in many ways. For example, students could buy a home-cooked dinner at the O Club for only $2.50 during the work week.

Because the social life of Enid couldn't match that in the big cities, we became more like family. Friday nights were special. Pilots met at the O Club to swap stories of their escapades that week. We'd have a few drinks and dance to live music or DJs until one in the morning.

I bought a small house on the edge of Enid. My yard backed up to a wheat field. With no city lights nearby, the sky was so dark that I could easily see the Milky Way. I enjoyed the view of the night sky so much that I bought two telescopes. I subscribed to *Astronomy* magazine and joined an astronomy book club. I learned the locations of all the stellar constellations and discovered that no matter where I was in the world, the sky was familiar and comforting. I had no formal courses in astronomy, so I became curious about the origin, nature, and possible destination of the universe.

I would wonder about the future.

Perhaps many generations from now, pilots will be training to go to places in the Milky Way galaxy, flying in ships yet to be imagined. How could I, an Air Force pilot in the age of the first Star Wars *movies, be a part of that future?*

Chapter 5

BE CAREFUL WHAT YOU ASK FOR:

OPERATIONAL PILOT

In 1982, I was twenty-five years old. It was time to plan the next part of my Air Force career.

To achieve my ultimate goal of becoming an astronaut pilot, I knew I'd need to graduate from the Air Force Test Pilot School (TPS) at Edwards Air Force Base. Even if I wasn't selected as an astronaut, being a test pilot would still be a dream job. I could easily spend the rest of my life flying the newest and fastest jets!

I needed to be realistic first. I had many hurdles to clear before I could even think about applying to TPS. The first was that a pilot needed at least one year as an aircraft commander in an operational aircraft, as well as one thousand hours total flying time.

I had already earned the one thousand hours while I was at Vance by flying day and night in the T-38. I flew every chance I got, including weekend cross-country flights with students from other sections. If the flight scheduler needed just one more pilot, I was there. They called me the squadron "time hog," meaning I logged the maximum flying hours allowed each month. Other instructors often grunted like pigs when they passed me in the hallway.

I loved this good-natured ribbing. Many of the married pilots, with responsibilities at home, were actually happy for me to fly the extra sorties. Being young and single, I just preferred to be in an airplane.

With the stick-time qualification met, my next task was to find the right operational aircraft. A trainer like the T-38 is not considered "operational," as it is simply a flying classroom. A pilot needs to fly an aircraft designated as a *major weapons system*, performing missions like reconnaissance, refueling, bombing, or transportation. As much as I loved the T-38, I needed to move on to another plane.

Air Force pilots fill out a "dream sheet" about six to eight months before changing assignments, listing in descending order the planes you prefer to fly next. Having flown the sleek T-38 and loving the Century Series planes, I yearned to fly a fighter. However, there was a legal obstacle on women pilots that the men did not have to overcome. The 1948 law that enabled women to pursue careers in the military also contained a "combat exclusion policy." It specifically prohibited women from serving on aircraft that might engage in combat. I had hoped, in vain, that the policy would change during my four years at Vance. The Air Force didn't rescind the policy until 1993, several years after I became an astronaut.

The F-15 was certainly out of the question as a first choice. I submitted my dream sheet with some other off-limits choices that might possibly be acceptable. First, I requested the F-106, an interceptor aircraft scheduled for replacement in active service by the F-15. My second choice was the RF-4, a reconnaissance version of the fighter-bomber that was a workhorse in Vietnam. My third option was a long shot, the A-10 Warthog, a close air support combat aircraft. Requesting the A-10 was unrealistic, as its operational mission was as close to combat as the Air Force gets, but there was no risk in my asking.

Shortly after I submitted my dream sheet, I called the assignment manager at the personnel center in San Antonio to see what I could finagle. He bluntly told me: "Captain Collins, you need to send me a form I can work with. You know I cannot assign you to any of these aircraft."

On the back of the form, I had listed every airplane I could think of. Combat restrictions took most of these off the table. I could fly air-refueling tankers like the KC-135, transport aircraft like the C-141, or a hospital aircraft like the C-9. Other options, such as administrative aircraft, advanced instructor training, or even the elite Thunderbirds aerial demonstration team, had limited availability and would not qualify me for TPS.

The C-141 Starlifter, a cargo-transport plane, was originally my twenty-first choice. I desperately wanted to pilot a plane that would fly me higher, faster, and farther. But considering the situation logically, the C-141 was a good compromise. It would provide me an opportunity to travel, as the C-141 carried troops and supplies all over the world. If I were assigned to Charleston AFB in South Carolina rather than the bases in California, I would be closer to my sister in Florida and my parents in New York.

I resubmitted my form with the C-141 as my first choice, although my heart wasn't in it. I requested assignment to Charleston.

The captain in San Antonio called me. "You can't fly at Charleston. They have airdrop missions, which you can't fly because of the combat restrictions."

"Why not?" I asked. "There are already women flying at Charleston."

"That's the problem. The schedulers can't put the women on the combat missions, so it ties their hands. They don't need any more women. Your choices are Travis AFB near Sacramento or Norton AFB near Los Angeles. You'll fit in there."

I fought his decision. I wrote letters. I pleaded with my commanders. No luck. I eventually picked Travis, as they had no airdrop missions at all. I'd be eligible to fly any and all missions available from that base.

There's an old saying in the Air Force: "Be careful what you ask for: you might get it!" But sometimes the greatest gift is an unanswered prayer. At Travis in 1983, I met the man who would become my husband.

I expended so much effort in fighting the assignment, but I didn't understand what was good for me! The Air Force sent me where I needed to be. The C-141 mission also prepared me for my eventual astronaut role much better than if I had been in single-seat fighters.

Double Diegos and Exotic Missions

I arrived at Travis Air Force Base in Fairfield, California, in March 1983, after a two-month copilot training course at Altus AFB in Oklahoma. It felt like a demotion and was a tough blow to my ego to transition to a copilot position after being an instructor and assessing other pilots on check rides. I vowed to upgrade to aircraft commander as soon as possible.

I quickly discovered the C-141 was much more complicated than the T-38. I realized that if I rushed too fast to upgrade, I might miss out on important experiences and risk the safety of the mission and crew because I didn't know the nuances of the aircraft and its systems.

Acting with humility proved to be an advantage. I challenged myself to learn the airplane better than anyone else knew it. I studied all of the manuals on engines, electrical systems, fuel systems, hydraulics, procedures, rules, and the responsibilities of the various crew members.

The C-141 had a large crew—commander, copilot, flight engineers, loadmasters, flight nurses, couriers, and others on board. Coordinating all these roles was a valuable learning experience. Our mission was to carry people and cargo all over the world for America's armed services. Many of these flights delivered humanitarian aid, and many supported military exercises.

I came into the C-141 world two years into Ronald Reagan's presidency. His administration worked to rebuild a strong military. While I was still instructing in T-38s in 1981, our classes grew from forty to sixty students to bolster the pilot force to the strength needed to implement the administration's policies. Now, two years later, C-141 crews significantly expanded their flying hours to support military exercises around the world. I was traveling about 80 percent of the time.

One of our domestic missions was the "Missile Run." The Air Force deployed Minuteman missiles at various bases around the upper Midwest, and those missiles needed to come back to the depot at Hill AFB in Utah for routine refurbishment. The missile would be prepared at the base and its nuclear warhead removed. Our loadmaster would place the missile body into our cargo hold, and we'd fly the missile to Hill. Then we'd load a refurbished missile and take it to its new home base. Repeating the process a couple of times per mission, we were usually on travel for five days at a stretch. I loved that when we'd get a day off at Hill, I could ski in the glorious mountains east of the base.

One of our routine itineraries was the "Double Diego" mission, with long hauls every day: Fly from Travis to Hickham AFB in Hawaii. Fly from there to Andersen AFB in Guam. Fly to Clark Air Base in the Philippines. Fly to the small Indian Ocean island of Diego Garcia, south of the equator. Fly back to Clark, back to Diego, and back to Clark, occasionally stopping in Singapore. Fly to Kadena AFB in Okinawa. Fly to Yokota Air Base near Tokyo. Fly to Elmendorf AFB in Alaska. Fly home to Travis. We would get at least one day off to rest and be a tourist at one of the locations. The overall mission typically lasted fourteen days and covered nearly thirty-five thousand statute miles. I would come home to an overgrown lawn, a stuffed mailbox, and a squadron inbox full of paperwork. We rarely used our allotted three days off to decompress between missions, as there was always more work to be done.

Another favorite route was the "Coral Run." We transported supplies to bases in beautiful tropical islands like Kwajalein, Wake, Johnson, and Midway.

Still another exciting trip was the "Reforger" or "Autumn Forge" exercise, which simulated the deployment of troops to Europe in the event of another World War II–type scenario. Our flights took us to England, Germany, Spain, and various other NATO countries. One time, as I was flying from Italy to Turkey to support an exercise, an angry Greek air traffic controller told me that I was off course and that he was going to "violate" me—meaning he would report me to the authorities for drifting over his country in violation of aviation notices. I found out later that he was just hassling me because of the Greek and Turkish dispute over Cyprus.

On one dark, moonless night, while following the air traffic controller's headings over Turkey, we sensed we were being vectored toward a mountain. I had to ask to change the approach in order to avoid my aircraft becoming shipwrecked like Noah's Ark on the side of Mt. Ararat. Nevertheless, the Turkish landscapes were absolutely stunning, almost a dream. The next morning, we flew around steep, barren, bright-tan mountains. As the sun rose over the wild terrain, I felt like I was flying above another planet.

I loved flying to Gander, Newfoundland, and Goose Bay, Labrador, in Canada. The winter air was thin and cold. Vast stretches of uninhabited tundra, lakes, and islands seemed ancient and unblemished. I was amazed at how clean and clear the air was in Iceland, a land that one could easily mistake for the surface of Mars. And I'll never forget how stunning it was to fly down the fjords in Greenland.

After flying extravagant and exciting trips like these, I wondered: *Why did I ever want to fly fighters?*

Culture Change

About a year into my assignment at Travis, I had just finished commanding the monthly flag-down ceremony, and my wing commander (the top boss on base) called me into his office. *This is getting to be a routine,* I thought, *senior commanders calling on the women pilots! Maybe they're just curious to see what we're like?*

Colonel Woods, who was an intelligent and sincere person, talked to me for quite some time while I stood at attention. "I see that women are doing well in their flying roles," he said, "but I'm not sure about the future of women in flying unit command positions. I need my wife to do this job with me. She fills in for me, especially in areas of family needs on base. She works harder than I do. The families need the commander's wife."

Although his words somewhat surprised me, I tried to hear his concerns, understand where he was coming from, and be open-minded about his perspective. After all, I had never been a commander at his level. He accepted women as pilots, but he foresaw challenges for women as they moved up the leadership ranks in the Air Force. I hoped he wasn't implying that a woman couldn't achieve a senior leadership position just because she wouldn't have a "wife" to oversee the human side of caring for people.

My gut told me it would work out okay when the situation did eventually arise. Surely, there will be female commanders and their husbands, vice commanders' wives, and spouses of many other commanders in the chain of command. There will be many other women in the ranks. *We can all support one another,* I thought. *So many different family situations are possible, and we will adapt.*

Culture change is messy and takes a long time. It's impossible to foresee all of the bumps in the road that will arise once a process is set in motion. We in the Air Force still had much to learn and many more issues to iron out, as I would soon discover.

Combat!

On October 23, 1983, a terrorist drove a truck bomb into a barracks in Beirut, Lebanon, and killed 241 US Marines. Our country was shocked and dismayed at such a horrible loss in this single devastating act of violence.

I received orders that day to fly a mission to take about sixty Marines from Cherry Point, North Carolina, to Beirut—an incredibly important and deeply sad mission. We flew from Travis to Cherry Point and then waited on the ground for eight hours while the logistics were straightened out. We could only load enough fuel to make it to Rhein-Main Air Base in Frankfurt, Germany. When we arrived at Frankfurt, our crew had hit the daily workload limit of eighteen hours. Rules required that we go into a rest period. Another crew hopped on our C-141 and flew the rest of the mission while we slept. When that crew returned to Frankfurt, they told us they had performed a blackout landing in Beirut—no lights on the aircraft, no lights on the runway.

We flew the same C-141 back to Dover AFB, Delaware, and were told to report to the command post for further instructions. Thinking we would have another mission to the Middle East, we were astonished when the officer told us that the United States had just invaded Grenada, a tiny island in the Caribbean off the coast of Venezuela. Our orders were to fly to Pope AFB, North Carolina, adjacent to Ft. Bragg, home of the Army's elite 82nd Airborne Division.

After the quick flight and minimum crew rest, we reported to the command post at Pope for our mission briefing. On the bus on the way to the meeting, we encountered another crew who had just returned from Grenada. The aircraft commander, Jeff Brake, was a fellow student of mine at Vance. I deeply respected Jeff, because he was the only student who had treated me as an equal by passing me the ball in our mandatory basketball games at Vance. (Back then, the women student pilots had to play basketball with the men, which resulted in many elbows to the ribs without any benefit to the score.) Jeff was ranting about how they had airdropped paratroopers from the 82nd Airborne into Grenada while being fired upon. The C-141 has no guns and no defensive capability, so Jeff had flown quite a daring mission.

I sat and listened to the scenario in the briefing room. We would be heading into a hot combat zone, with active fighting against Cuban soldiers still underway on the ground. Having heard the account of Jeff's previous mission, I asked the command post what we would do if we were intercepted. The briefer said, "Go into a cloud." The guy was clueless. Modern antiaircraft missiles are heat-seeking and radar-seeking, so what good would a cloud do for us?

My aircraft commander was Glenn Chinn, a calm and focused pilot. After listening to the briefing, he readily accepted the mission.

We knew the risks. Off we went.

We loaded our aircraft with 140 soldiers and their support equipment. As we started the engines, our loadmaster called to us over the intercom, "We can't go. We only have 110 oxygen masks." We made several calls to the over-tasked command post for more masks and were eventually told, "We are out of masks. Go anyway."

We knew the rules for safe operation of the aircraft. However, we learned that in a time of war, sometimes you throw the rule book out the window. If we lost cabin pressure on the flight, the brave soldiers in the cargo hold would just have to share their masks.

Since there would be no refueling on the ground in Grenada, we had to load up heavy with just enough gas to fly 1,927 miles, offload the troops on the runway without shutting down the engines, and then fly about another 1,838 miles back to Charleston. It would be tight—right at our maximum range.

The flight south was quiet and without incident. We learned on the way that the previous missions had secured the airfield in Grenada. Regardless, for additional safety, we flew our final approach over the water.

When we landed, I was shocked to see the war zone firsthand. It looked like a movie set, with barbed wire, sandbags, soldiers, and action everywhere. We rolled to a stop. The troops in the cargo hold started jumping to psych themselves up, and they shook the aircraft like a wild thunderstorm. We opened the cargo hold doors, and the soldiers rushed out.

We intended to turn around and take off immediately, heading back out over the water. Instead, we received instructions to taxi to a nearby ramp to pick up passengers. "We don't expect passengers, thank you," Glenn radioed. The onsite commander replied that the Cubans had captured and recently released thirty-six medical students who were en route to the airport with their families and needed a ride back to the United States. We quickly loaded them on board and headed out.

During the flight home, the young daughter of one of the students drew me a sketch of our C-141 with the words "Thank you, pilots." The students came up to the cockpit, one by one, saying how grateful they were to President Reagan and the United States for rescuing them. The Cubans had released them just the day before, and you could tell how frightened they had been. The flight home was otherwise quiet. We had to climb to 43,000 feet—well above our usual maximum altitude—to stretch our fuel and make it back to Charleston.

The base commander, the mayor, the city council, and local media met us when we landed. Some of the medical students kissed the ground when they got off the plane. Glenn and I watched all this through our cockpit windows. That night, we saw some of the students interviewed on national TV.

It was a successful rescue by the United States—a shining example of President Reagan's policy of "peace through strength."

The Air Force gave me combat pay for flying that mission, something like $75. They awarded us an expeditionary medal and a commendation medal. Some aircrews received an air medal. Women were receiving combat awards, although we were forbidden to participate in combat. Imagine that!

When the fighting starts, there's no time to pull women out of crews. We were all in it together.

Passing the Hours

Since our crews spent so much time in one another's company, we had plenty of time to joke around. A crew member once found a huge pair of women's underwear in a drawer in his room on a military base. This underwear passed from luggage to duffel bag throughout the trip, with no one knowing who would eventually be taking it home—hopefully not to be found by someone's wife! Another time I ended up being the butt of a joke, and in retaliation I put a plastic spider in the flight engineer's coffee cup, with tremendous results. We read funny stories to one another to burn the long hours flying over open ocean that could often be tedious—though not for me.

I was rarely bored and almost never slept on long flights. The view was always too beautiful. There were stories to share or maps to study, colorful sunrises and sunsets, flashing lightning bolts from distant thunderstorms, bright white billowing cumulus clouds, nighttime sparkles of St. Elmo's Fire, and tiny island atolls so small that no map recorded them. If you've never witnessed the world unfolding before you from the cockpit of an airplane, you wouldn't believe what you're missing.

Our travels weren't always fun or beautiful. While the daytime shopping around the US bases in the Philippines was a great cultural experience, the nighttime adventures disturbed me. There were many seedy bars and other venues for bad behavior. The senior woman aircraft commander in our squadron, Janet Kunchew, took me to some of those places because she said I needed to know what my enlisted troops were doing during their crew rest period. I have a particularly sad memory of a little girl, who couldn't have been more than three years old, selling flowers by herself outside a bar. I wanted to take her home with me, or at least back to the safety and security of the base. Janet told me to lay off, as the girl's mother was one of the exotic dancers inside the bar.

After observing the tough living conditions in many countries in the Americas, Asia, Africa, and some places in Europe, I learned how fortunate we are to be in the United States. Here, even those who live in poverty have access to food and basic services. We can be grateful for the opportunities afforded to US citizens. Despite growing up in government-subsidized housing, I was blessed to be able to attend schools, have an opportunity to improve myself, and become part of the American dream. This is not something our fellow humans can aspire to in some other areas of the world.

I participated in many memorable missions, some of which I'm not at liberty to discuss, and the most rewarding by far were those helping the needy. Most Americans are not aware of the humanitarian missions flown by the US military. Our C-141s supported digging of water wells in Central America and relief efforts after hurricanes and volcanic eruptions. These types of missions are still a high priority today. They help struggling communities, are great training for our aircrews, and promote international relations and diplomacy.

I could often guess what my next mission was going to be just by watching the international news. If there was a hot spot in the world, we would surely be called there to help out. I was proud and humbled to represent the United States on these missions.

Growing in the Role

NASA issued a call for astronaut applications in 1983, for the third class of space shuttle astronauts. I met the criteria for eligibility, if I chose to apply. I filled out the application paperwork and completed it all except for the letter of recommendation from the wing commander. I was still new on the base, and I hadn't yet established myself. I was nervous about going to Colonel Woods and asking him for the letter. I thought he might just laugh at me and say, "You just arrived! Why do you want to leave?" And so, I talked myself out of submitting an application.

I concentrated instead on learning while advancing in my role and responsibilities. After nine months as a copilot, I received my upgrade to first pilot in November 1983. Next, I returned to Altus AFB for aircraft commander school and air refueling training. My full upgrade to aircraft commander came in June 1984, and then I began working on becoming an instructor pilot.

I realized that flying the C-141 was excellent training for someone who might someday command a space shuttle mission. Our flight crews were analogous to space shuttle crews. Both had four to eight crew members with a mix of pilots, flight engineers, loadmasters (mission specialists in the shuttle world), flight nurses, and other specialists, all of whom the commander had to lead as a high-performing team. Both vehicles flew diverse payloads for a wide array of customers.

The aircraft commander had similar responsibilities to a shuttle commander's in caring for the vehicle and the crew while responding to complex situations. I saw many occasions in which the C-141 commander was far from base or out of radio contact and had to make critical, life-or-death decisions for his crew without communicating to the command post.

Once, as we were flying over the Arctic, we had a fire in our throttle quadrant—the pilot's controls for the plane's engines. Another time, our flight engineer called to us, "The number two engine is putting out flames!"

The aircraft commander had to be comfortable with many leadership styles and know which ones were appropriate for a time-critical situation, using only his or her experience and judgment. I quickly learned to trust my enlisted crew, as many of them had been doing their jobs for more than a decade. They in turn respected aircraft commanders who listened to them.

One of my most valuable leadership lessons was "cockpit resource management" or "crew resource management" (CRM). Developed by NASA and the National Transportation Safety Board in the late 1970s, CRM provides a structured way for flight crews to work through problem situations through better leadership, communication, and decision making.

The basic tenets are: know your job and do your job; be aware of what the other crew members are doing; and communicate clearly and check for understanding. It seems simple on the surface, but you would be amazed at how quickly a situation can become deadly if any of those steps are missing.

The classic example is the story of an aircrew that died when their plane flew into a mountain. Everyone in the cockpit focused on a faulty landing gear light during final approach. The pilot became so distracted that he didn't realize there was a mountain looming in front of the plane. The pilot should have done his job by concentrating on flying the aircraft and knowing where the aircraft was, while letting the engineer work on the light. This terrible tragedy could have easily been prevented. Commercial airlines, the military, and

NASA still teach CRM today, and the accident rate has decreased significantly as a result.

And like so many other things I learned as a pilot, the CRM mindset made me a better partner, parent, colleague, and leader. There are so many times when it is easy to fall into the trap of assuming that everyone in a group shares the same view of a situation or the path forward. Knowing your role, looking out for one another, and being comfortable with checking to be sure you're all on the same page will save time and prevent wasted effort.

Not Just Dating My Airplane

Every officer had a desk job when they were not flying. Mine was in the squadron's Executive Office, writing performance reports for officers and airmen. As is the case with a lot of those administrative roles, I was put in the job because I was available, not because I was good at it!

I was picking up papers in the scheduling office one day in August 1983, when a guy came in wearing golf clothes. The scheduling officer introduced me to Pat Youngs. I had heard his name around the base—a C-141 pilot who was also a hotshot golfer—but I had never met him. We exchanged a few pleasantries, then I walked back to my office.

He followed me in and asked, "Do you want to go out to dinner tonight?"

I felt astonished that he would be so bold as to ask me out when we had only just met. Instead of acting affronted, I found myself saying, "Sure, I'll go out to dinner."

He said he'd pick me up at six o'clock.

I went home after work and changed. Six o'clock came and went. Then six-thirty. I was hungry and irritated, so I ate an entire tub of ice cream and some other junk food. Pat finally showed up more than an hour late. We drove to a nice restaurant near UC-Davis. I could only eat a salad, having devoured all that ice cream before Pat arrived.

Pat was honest with me about his tardiness—he felt the need to take advantage of every hour of daylight on the golf course. Lesson learned: if I was going to date this guy, I'd better become a golfer!

Like me, Pat was a C-141 pilot, and he left on a Double Diego trip the day after our first dinner. When he came back to Travis two weeks later, he mentioned that he was going to shoot baskets at an outdoor court. This time, I surprised *him* by inviting myself to come along. That was our second date. The next day, I headed out on a mission.

It's hard to date someone regularly if you are both active-duty military pilots, because you are always out flying. If you're lucky, you might see each other once or twice a month.

Pat and I dated for four years before we married. We were not in any hurry in our relationship. Pat had his golf game to perfect, and I had my career ambitions of getting into Air Force Test Pilot School and eventually NASA. At our young age, we felt like we had our whole lives ahead of us.

Chapter 6

AIR FORCE ACADEMY

With several years of flying and command experience in the C-141, I had passed another of the qualification thresholds for the Air Force Test Pilot School, and I was becoming a more competitive prospect for NASA's astronaut program. Now, I needed something to make me even more attractive as a candidate for my dream positions.

In the pre-internet days of the 1980s, I learned about potential opportunities and their requirements by reading the *Air Force Times*, talking to the Air Force Personnel Center, and researching regulations. I had to keep an eye out for openings and be ready to apply when they became available. Nobody sought you out to tell you to apply. Your superior *might*—but only if your boss considered your career potential and the good you could do for the Air Force rather than how much he needed you to stay and support him.

Flying for the Thunderbirds, the Air Force's precision demonstration squadron, seemed like an appealing choice. However, there were several strikes against it. First, NASA wanted its shuttle pilots and commanders to have test pilot experience. Being a member of the Thunderbirds wouldn't count toward that requirement.

After a horrible accident in 1982, the Thunderbirds changed from flying T-38s to the new F-16 Falcon fighter. Because the F-16 was a combat aircraft, policy forbade women from flying it, even if just for air shows. They justified the exclusion by saying the Thunderbird pilots could go into combat on a moment's notice.

My thoughts turned next to the Air Force Academy in Colorado Springs. I could apply for a position as a professor. This could be an opportunity to satisfy my early ambition to teach math or science.

I contacted the academy, and they agreed to bring me on as an assistant professor. First, though, I would have to earn a master's degree, which the academy agreed to pay for. This was looking better and better, as NASA strongly prefers astronaut candidates with advanced degrees.

The academy connected me with a program at Stanford University through the Air Force Institute of Technology. I chose Stanford's degree program in operations research (OR), a combination of industrial engineering, modeling, and computer programming. It involves decision making in large systemic problems by using analytical techniques—a perfect fit for me. It would qualify me to teach, and it would enhance my value to the Air Force. Additionally, I would take part in one of my favorite activities—problem solving.

After seven years of flying high-performance jets, attending a university was more difficult than I expected. I thought I would fall back into academic culture easily and happily. However, I soon realized my heart was empty. I wanted to get back into operations, *now*. Stanford was a beautiful location and an intellectually challenging environment, but how could I possibly survive here for eighteen months? Not only did I miss the operational world, but I saw my chances for test pilot school and the astronaut program possibly slipping away if I was too old to apply.

I quickly replanned my remaining academic terms, overloading courses so I could complete the program six months early. I finished my master's degree in eleven months. My grades were not as good as those of my undergraduate years, but I passed, and I earned my degree. That's all that mattered.

I was headed back to the Air Force. I couldn't wait to get to Colorado Springs.

Teaching at the Air Force Academy

One other important factor compelled me to move to Colorado as quickly as possible. I was still dating Pat, and we wanted to continue seeing each other. Independently of my decision to teach at the academy, Pat received an appointment to their athletic department as a golf coach. The timing was sheer luck. He was about to be transferred to Altus, Oklahoma, but just in time, the three-star general at the academy offered Pat the coaching position. Pat had graduated from the academy in 1980, so this was familiar turf to him.

I arrived three weeks after he did, in August 1986, to begin my stint in the math department. We purchased new homes three blocks away from each other in the Gleneagle community, due east of campus.

I taught five sections of precalculus to freshman cadets. Most of my students were football players who had attended Air Force prep schools and needed some extra help. This was a great way to start my teaching career. I enjoyed their questions, and I learned to put myself in their mindset to understand how and why they might struggle with math. Each subsequent semester, I taught the follow-on calculus course: differential calculus, integral calculus, then multivariate calculus. I became a course director in linear algebra—a basic course in analytical problem solving, which included computer programming.

I had many growth experiences of my own during my three years as an assistant professor and classroom teacher. First, I have always been terribly fearful of public speaking. I found the best way to overcome this fear is to become a classroom teacher. Once you thoroughly know your subject and are passionate about passing along your knowledge, you can forget about yourself as a speaker and focus on the points you want to make.

I also recognized the importance of training young minds to solve problems. While mathematics is black and white—there are right and wrong answers—the true benefit of learning math is the problem-solving methods. Mathematics is a mindset. Define the problem, name the variables, label the constraints, set up a list of your "tools" (such as previously proved elements), put your steps in order, and so on. This is such an important part of why we learn math, at all ages and grades.

When I wasn't teaching math, I instructed student pilots. The flying mission at the AFA in general is to prepare pilots and leaders of the future. Our squadron's specific mission was to be a screening program similar to the one I attended at Hondo eight years earlier. My primary job was to evaluate whether my students could handle the rigors of Undergraduate Pilot Training, which they would attend after their Academy graduation. The secondary mission, which was just as important in my opinion, was to teach them the flying basics.

We flew the single-engine propeller-driven T-41 Mescalero, a souped-up version of the Cessna 172, which is still the most commercially successful airplane in history. The T-41 had a 210-horsepower engine to give it the performance required to fly in the thin air at the 6,500-foot altitude of the AFA airfield.

The T-41 was the perfect airplane for this mission. Its simple, safe design allowed the students to focus on basic flying skills, navigation, radio communication, and situational awareness. In addition to getting a break from the classroom and grading papers, instructors enjoyed the side benefit of viewing the gorgeous Colorado landscape and stunning mountains from a higher perspective.

Marriage and a Military Career

For the first eight years of my Air Force career, I was "married to my airplane." This mindset served me well while I learned the ropes and developed the discipline that I'd need to be successful—and survive—as a pilot. I wanted to spend as much time as possible in airplanes, and I wanted to see how far I could advance in rank while still staying in the cockpit. I used to believe there was no time for marriage and a family while I was flying, and flying would be my entire life.

But after dating Pat for more than three years, I realized I might actually be able to marry him and still pursue my life's dreams.

Pat proposed to me six months after arriving at the academy. We were married four months after that, on August 1, 1987. Our lives did not change much. Pat sold his house and moved into mine. Other than that, we went back to work.

Our biggest change was that we could no longer fly together. I had enjoyed the few times we flew operational and training missions together. However, one week after we tied the knot, we were walking out to the T-41 for an instructor proficiency flight. The squadron commander ran after us and caught up to us halfway to the airplane.

"Where do you two think you are going?"

"To fly, sir."

"Oh no, you're not! Air Force Regulation 60-16: 'Family members are not allowed to fly operationally in the same aircraft.'"

While Pat headed back to his car, I went into the squadron office and looked up the regulation. The commander was correct. The Air Force created the rule decades ago, after losing an entire airplane full of people during a spouse familiarization flight, leaving behind several orphaned children who lost both parents in that single accident.

Married couples in the military face a difficult challenge because they might receive different duty station assignments. The rule is "The needs of the military come first." Since we married one year into a three-year assignment, we still had time to work the system. For now, I felt confident that Pat supported my career goals, and I supported his. My passion was flight—and shooting for spaceflight—while he was still perfecting his golf game. For the time being, we could make it work.

We knew that eventually our luck might run out, and the Air Force would send us to different locations. That was not how we wanted to live our married life. Staying with the Air Force was the only way I could achieve my life's dreams. Pat wanted to fly with a major airline. He could live anywhere I was stationed and commute to his base, which gave him the flexibility to follow me. He'd have time to compete in golf, as well. He hired on with Delta Air Lines and flew with them for thirty-one years. Our arrangement worked beautifully.

Planning for the Big Play

Not long after my arrival at the AFA in 1986, I began planning my next career move: getting into the elite Air Force Test Pilot School. Oh, how I dreamed of becoming a test pilot! I wanted to fly the newest aircraft, operate the latest systems, develop the mission rules for the future, and work with the most elite pilots around. I wanted to fly the same types of missions as the legendary Chuck Yeager and Scott Crossfield. They were my heroes: confident, focused, daring, and the first in their field. They were leaders in the future of aviation and space. I wanted to follow them, so to speak, and then go beyond. It could be the pinnacle of my Air Force career.

I waited for the announcement in the *Air Force Times* and applied during the annual admissions window, although it was during my first year of teaching. Completing the application for test pilot school wasn't difficult, but meeting the requirements was. I had been diligently working for years to build the requisite skills. Bachelor of Science degree in math, physics, or engineering: check. One thousand hours flying time: check. Twelve months of experience as an aircraft commander in a major weapon system: check. My weapon system, the C-141, would qualify me as a test pilot for the future C-17 Globemaster III. The C-17 was under development to replace the C-141, and I was ready to fly it! The astronaut program was also looming as a possibility, but I wasn't yet at the point where I felt I was competitive. So, I focused my energies on the test pilot school.

After submitting my application, I held my breath awaiting the announcement of the list of new students.

My name was not on the list. I felt crushed.

I waited a few days for the news to sink in. Then I called the Personnel Center and asked, as diplomatically as I could, what happened.

"Your application never went to the selection board," Major Jackson told me.

I was stunned. "Why?"

"Air Force policy is that no one can be transferred until they have two years 'time on station.' You have to spend another year at the academy."

That was hard news to hear, but at least the rejection was just due to policy—nothing to do with *me* or my qualifications. Rationally, I could see that another year teaching would allow time to prepare. I still enjoyed classroom teaching, so I wasn't stuck in an unpleasant assignment.

I applied again a year later. This time I followed the application more closely, and I called Major Jackson before the board met. He surprised me by saying, "I'm sorry, but you can't meet the selection board this year."

Once again, an Air Force policy stood in my way. He explained, "You are on a 'directed duty assignment.' You must spend three years at the AFA before you can be transferred."

That policy requires officers to pay back their college tuition by serving in the duty assignment that funded their degree program. Since the AFA paid for me to attend Stanford, I had to work for them for three years in compensation for their investment in my development.

It made perfect sense to me, but I hadn't remembered that restriction. I was thankful Major Jackson cared enough to let me know before the bomb was dropped. Nonetheless, I was upset that my paperwork again went straight into a filing cabinet.

I tried diplomacy, everything short of pleading. I asked if I could apply for a waiver to the policy. "After all, I'll still be using my degree at TPS."

"No," he said. "No waivers allowed."

And now there was another administrative glitch looming. Rules said that you couldn't apply to test pilot school if you'd served in the Air Force longer than ten years. I would pass that mark before next year's application process.

I told him I was in a bind. "Either I have to apply for a waiver to the directed duty assignment, or I have to apply for a waiver to the ten-year time-in-service limit."

Major Jackson told me the time-in-service waiver was workable. He even offered to help me with it the next year. I was leery. Much can happen in a year. The Air Force constantly changes their policies. And who knew if Major Jackson would even be there next year or if he would have the authority to help me?

I reminded him of the old joke, "How do you know when the personnel officer is lying? His lips are moving."

He laughed. "Don't worry, Captain Collins. We will get you a time-in-service waiver next year."

I look back in wonder at how brash, bold, or unreasonable I often was in pushing for things for myself. When I think about my audacity in protesting decisions about my duty locations early in my career, I shake my head. I always felt like I was fighting the system or the timing was wrong. However, test pilot school *was* the right fit for me, whether or not I went on to NASA's astronaut program. I didn't want to lose out on the chance to realize a lifelong dream just because of an arbitrary policy—or because I might be technically too old for the assignment.

I wasn't going to give up on this one.

The year I missed the selection board the second time, the school chose its first woman pilot. Jackie Parker had a similar background to mine. Although several women flight engineers had attended test pilot school before, Jackie became the first woman pilot at the school.

When the list for the 1988 test pilot classes came out, I noticed the pilots were younger than in prior years. I was concerned that the Air Force was boxing in their pilots. How many pilots can earn all the prerequisites prior to ten years of service? If you include time earning an advanced degree, and time spent in nonoperational assignments like Air Training Command, you eliminate many potentially qualified people. I tried to rationalize my situation, but these decisions happen at a much higher level than Major Jackson's and mine.

Waiting until the following year was agonizing. In 1989, the third time I applied, Major Jackson came through with the time-in-service waiver, and I was accepted. What a relief! After all the false starts, finagling, complaining, arm-twisting, worrying, and waiting, I finally made it! I would become the second woman pilot to attend the Air Force Test Pilot School.

Looking back, I see that things actually worked out superbly for me. I finished my tour of duty at the Air Force Academy, pinning on my new rank of major during my final month. I taught and attended courses that improved my understanding of analytical modeling, computing, aeronautics, and astronautics. I earned another master's degree in my spare time, this one in space systems management. Best of all, Pat and I married after our first year of teaching!

And now, I would be heading to Edwards Air Force Base in California for every pilot's dream job.

Chapter 7

TEST PILOT SCHOOL

As soon as the Air Force Test Pilot School announced my acceptance, I was mentally and emotionally ready to start training. I was laser-focused on learning the process and the methods of succeeding in this difficult career. My education began even before I officially became a student.

All newly accepted pilot applicants must pass a series of five flights at Edwards AFB several months before classes start. This "Five Flight Eval" program runs an entire week, and all five flights are in the T-38. Initially, I thought the school's leadership wanted to get an idea of each pilot's ability to fly basic maneuvers, though that seemed somewhat strange to me. All the selected pilots were obviously capable, and everyone had previously flown the T-38 in Undergraduate Pilot Training. *So why do we need to do this?* Then I guessed they just wanted to get to know us and see our performance under pressure. After all, there was a catch: the school could still reject you if they saw something they didn't like.

By the time I finished the first of the five flights, I understood the rationale. I was asked to plan a flight with several flight test maneuvers and to verbalize my intent and observations while actually flying each one. My communications skills were an essential element of this evaluation. Additionally, I learned that being a test pilot requires a more complex set of skills than merely flying the airplane. We had to prove that we had the ability to connect the test objectives to the handling of the aircraft.

This was about so much more than mere piloting skills. Think of it as precisely flying the airplane, evaluating its handling qualities, and monitoring detailed performance data, while simultaneously conveying the larger context about how you, the airplane, its systems, and the mission relate to one another.

In essence, test pilots *write* the rules for an aircraft. Operational pilots *follow* those rules. Test pilots must be able to think creatively. They must determine and evaluate the limits of an airplane and its systems for functioning successfully and safely in the operational world. As technology is rapidly advancing, a test pilot must be able to adapt to changes and logically determine how technology can enhance the airplane and the mission—or how it might detract from safe performance.

It was an eye-opening week. After my Monday-through-Friday evaluation by the school's instructor pilots, I felt pretty good about my performance. I was certain I would fit in with the culture and mission of flight test.

Just prior to checking out to return home that Friday afternoon, I encountered the school's commandant, Colonel Mike Kostelnik. People who didn't know him well feared him, since he came across as a tough guy. He liked to remind people that at one point he was the youngest colonel in the Air Force! That achievement was possible only with early promotions for every rank up to and including colonel.

Colonel Kostelnik recognized me as the one person in the incoming class with the rank of major, which by definition made me the senior officer, with the additional responsibilities of class leader. Without any welcome or introduction or wasting time with chit-chat, he confronted me point-blank, revealing what concerned him most about having a woman as the senior student.

He first words to me were "Will your husband run the wives' club?"

I assumed his question was a joke. I laughed and said I didn't know. His serious expression conveyed immediately that it was no joke. The welfare of wives and families on base was obviously of great concern, and he repeated the question.

Wow, this guy is living in the 1960s!

"Yes, sir, I will ask him!"

That was that. He walked away, knowing he'd conveyed the message he intended. I flew back home to Colorado Springs, confused about a dilemma I hadn't expected as the first woman class commander at the Air Force Test Pilot School. It reminded me of the lecture my squadron commander at Travis gave me about women in leadership roles.

I tried to put it out of my mind. *Stay focused, Eileen!* I told myself. *You passed your evaluation; you are going to be a test pilot!* Just being here was all that mattered.

When I returned home, I told Pat that the test pilot school commandant wanted to know if my husband would run the wives' club. Pat's quick wit is never challenged. He thought about it briefly and responded: "Yes, I will. We will have golf tournaments every Wednesday, and lingerie shows every Saturday."

Good grief, I thought. *This is unprecedented.*

What else am I in for?

Settling In at Edwards

Pat and I drove onto Edwards AFB through the north gate on a Sunday in June 1989. *This place reminds me of the gunfight at the O.K. Corral*, I thought. I also wondered, *Is this what the surface of Mars feels like?* My years of reading stories about test pilots had partially prepared me for the climate. I knew it would take some getting used to.

It turned out this was not an average day at Edwards. The base is in the high desert of California, about halfway between Los Angeles and Death Valley. While a summer day can routinely hit 100°F, there are not always forty-knot wind gusts and dust devils bouncing down the runway. The climate can usually be quite nice, and the beautiful smell of the clean desert air was refreshing. As time went on, I learned to love the landscape, with its Joshua trees, and the flat running trails. The endless wide-open blue sky was the ideal place for flight test.

Our first day of class was mostly spent socializing and getting to know one another. Class 89B, the second and final class of 1989, included fifteen pilots (one from India and one from Canada), six flight test engineers, and four navigators. Of the pilots, six were qualified in multiengine heavy aircraft like tankers, transports, and bombers. The rest were qualified in fighters. There was only one other woman in the class, a civilian flight engineer named Norma Taylor. We held a reception in the foosball game room, selected our seats in the classroom, and then attended a briefing in the theater.

Our new commandant, Colonel ET Pollock, delivered a classic speech. He bluntly told us that since we were now becoming test pilots and flight test navigators and engineers, we were out of the Air Force mainstream. He said, "You are in left field. Get used to it. The *real* Air Force has forgotten about you."

My class did not take this well. We designed our class patch with the motto LEFT FIELD.

As part of our indoctrination, we watched a test pilot recruiting video from the early 1960s titled *Roy and Helen*. This corny short showed the perfect Steve Canyon–type fighter pilot rolling into Edwards AFB with his perfect wife and two perfect kids in tow. They drove up to the school in their perfect 1950s car, family smiling, with Roy on his way to a perfect career. I'm still not sure whether the film was supposed to be a serious introduction or if it was a running joke.

We couldn't resist satirizing it, though. My classmate Jeff Smith and his wife, Denise, produced a 1980s version of the recruiting video, except reversing the roles in the perfect family and calling it *Pat and Eileen*. Eileen took the role of Roy, and Pat was in the role of Helen. We made hilarious scenes of Pat taking the class's twenty children to school and then burning dinner. Eileen displayed her dissatisfaction with his housekeeping and then went out to fly. It was a fun way to recognize that times were changing, and it lightened the mood as we began a stressful year.

I was still struggling with another tradition that was being challenged—figuring out what to do about the wives' club. The second-highest ranking person in our class was Harry Whiting, a talented, confident, and boisterous fighter pilot. I told him my predicament and asked whether his wife, Jackie, would take the job. Harry dutifully asked Jackie, and she agreed.

Even in 1989, the whole concept of a wives' club was already out of date. Maybe it made sense in the 1950s and 1960s, but certainly not now. I felt sure the wives (who we now refer to as spouses) didn't want someone "running" their activities. However, it wasn't my place to force a change in Air Force culture just to fit my own ideas of how our families should operate. I knew that time would pass and people would adapt.

For now, though, crisis averted.

Work Hard, Play Hard

All student test pilots lived in homes on Sharon Drive. Single (unaccompanied) students lived in nice, compact apartments called the bachelor officer quarters. I liked the idea of the student pilots all living in the same neighborhood. Many of the married couples had children, and this arrangement allowed families to support one another. It also built camaraderie.

Over the years, the senior class developed a tradition of indoctrinating the new junior class by playing some kind of prank. Having a student test pilot neighborhood made us easy targets.

Since I was the junior class leader, Pat and I hosted a party for my classmates and their spouses. I thought it would be a good opportunity for us get to know one another in a fun environment, and the spouses could meet the other students.

We had a wonderful time. We were playing a wild game of "spoons" when suddenly all the talking and yelling stopped. Something was violently shaking the house. I looked up at my ceiling and watched a crack propagate from one side to the other.

The terrifying crackling sound and vibration weren't from an earthquake. It was the senior class jumping up and down on the roof, the dreaded "roof stomping" that I'd heard stories about. They had some pretty big guys, and they were determined to welcome us in the worst possible way. Or maybe they just couldn't take the stress of the difficult year at TPS, and they needed to blow off steam. Or maybe they had lost all self-esteem and reverted to caveman mentality, or they wanted to show us who was in charge. Whatever it was, I wasn't impressed! These homes, built in the 1950s, had no attic. The flat roofs and ceilings weren't designed to hold any substantial weight. The guys in our class went outside and shooed the seniors away, and of course they immediately disappeared like a pack of guilty criminals.

I had just signed off on the paperwork accepting the house the previous day, so I was responsible for its condition. On Monday morning, I called the housing office and reported the damage. Little did I realize how much trouble this would cause. The housing office wasn't happy, so they called Colonel Pollock to complain about the behavior of the students.

Colonel Pollock summoned us all into the auditorium later that day and chewed out both classes. "You are professional student test pilots. You are responsible for your actions. You are responsible for US government property, whether it is an airplane or a house. Any damage you create as a class you are responsible to fix. But I *never* want to see any permanent damage."

Our class immediately took that on as a class motto. We added No Permanent Damage to our class patch. It was an attitude we could all support.

Now our challenge was how to get revenge on the senior class without causing any damage whatsoever. A few weeks later, after hosting a second party, we climbed on the roofs of each of the senior class members' homes and unplugged their swamp coolers—the desert versions of air-conditioning units. The poor senior student pilots each woke up in the middle of the night in a hot sweat.

That was the last incident involving the homes on Sharon Drive.

A month into our training, the senior class went on a two-week field trip. My class decided to execute another prank, which I believe was a work of art—a seriously sophisticated engineering project devised by some genius military minds. What could motivate military minds better than a golf course inside the school itself? And since the senior class was out of the state and their classroom was locked, why not put the golf course in their classroom?

B-52 Aircraft Commander Jeff Smith was the mastermind, and he assembled a crew of 89B classmates eager to avenge the roof-stomping incident. While I approved the idea, I reminded everyone, "Please: No Permanent Damage! Class Motto!"

It went like this: They received permission from the staff pilots and acquired the key to the senior class room. Then they mixed and applied concrete to construct a sidewalk and reinforced it with rebar. They borrowed a golf cart—which they had disassembled to get through both doors—and placed a ramp to get it up the stairs to the classroom. They brought in a child's pool for the water hazard and filled it with water. They "borrowed" a ball washer from the golf course (which they returned later), wheelbarrowed in hundreds of pounds of sand for the sand trap, and placed a tree. The wives covered every inch of the wall with poster paper, painted with palm trees and ocean views.

The senior class returned from their trip late at night—and boy, were they impressed. Even they had to agree that this was an engineering feat. It was successful, beautiful, useful for real practice, and we were able to keep it a secret from everyone except for those few who were in on the joke. We certainly earned mountains of respect from people all over the base. Most important, the senior class started to realize maybe they were dealing with a formidable enemy after all.

You are probably wondering: how did the students have time for this? Did they actually do any *real* work?

Any fun we had was part of the "work hard, play hard" attitude at TPS. The work by day was incredibly stressful, and often dangerous. The test pilot world needs people who are willing to take risks. You need people who are willing to fly maneuvers that have never been flown before. That kind of person tends to be more "creative" when it's time to decompress after a relentless week of studying and flying.

I don't think people did unsafe things that would hurt anybody. Despite the stories we enjoy telling, it wasn't like a big frat house. Our after-hours antics provided a way for hardworking people to blow off steam and build camaraderie. If you took it too far, you could be dismissed. That happened to one of the pilots in a previous class. Too many pranks caused him to lose his focus and wash out of the program.

The overall meat of TPS was serious and conservative. The top test pilots—such as Chuck Yeager, Scott Crossfield, Neil Armstrong, and John Young—were focused and professional, not hotshots. Deep down, we all hoped to emulate their example. Each of us wanted to be the best test pilot we could possibly be. Everyone wanted to get the highest grades on their flights, verbal presentations, written presentations, and tests.

We were so aware of our natural drive to be the best that from the outset, we made rules to keep the level of competition from becoming overwhelming. For example, classmate and U-2 pilot Rob "Skid" Rowe suggested on the first day that if you told anyone you got a perfect score on a test, you had to buy beer for the whole class. Otherwise, you needed to keep your grades to yourself. That got us off to a good start.

We were always working. We often would come home for dinner and go right back to the classroom. Maintaining work-life balance was nearly impossible. That's why they put all the married families together on Sharon Drive. The spouses needed to support each other, as everyone was going through the same challenges.

I wasn't about to waste any possible opportunity to get the most out of this experience by being unprepared. I was always studying, memorizing, or practicing. There was no time to follow the national or international news, but one event stands out in my memory. During a late night in the classroom, a janitor came in and said, "Turn on the news! They're knocking down the Berlin Wall!"

Learning to Be a Test Pilot

Whenever I start a new position, I habitually define the mission of the organization I work for and then determine my role to support the overall mission. At TPS, I had two missions. First was to become the best test pilot I could be. Second, I needed to perform my duties as class leader, to help those in my class achieve mission number one for themselves. The entire reason we were at Edwards AFB was to become the best test pilots, navigators, and engineers that we could possibly be. I tried to focus on that every single day.

The TPS curriculum was divided into three parts: Performance, Flying Qualities, and Flight Test Planning. The Performance segment came first, and it included flights primarily in the T-38 and F-4. I was assigned to the T-38. We flew maneuvers like "level accelerations" in instrumented aircraft, which were capable of transmitting performance data from the airplane's systems to the school's data room. We even flew high-speed "tower fly-bys" where engineers on the ground gathered flight data using techniques developed decades ago. Our training began with the basics and worked up to more sophisticated measures. Measurements such as airspeed, altitude, and pressures were downlinked many times a second.

The student had to make sense of the data by organizing it into *information*, usually graphs and charts, which connected the dots and explained the capabilities of the aircraft. It's not just "how fast can this plane go?" Rather, you might be developing information about what happens to engine pressures or the behavior of control surfaces as you accelerate. One of my core lessons learned was "You always need to know where your data are!" I carry that mindset every day.

The Flying Qualities phase primarily used the F-4, a fighter aircraft that saw extensive service in Vietnam in the 1960s and 1970s. Everyone complained about the F-4 and its many quirks that made it difficult to fly. The fact that it "handled like a dog," as most people said, made it an excellent instructional aircraft. The evaluations in this phase were more qualitative. We used the subjective Cooper-Harper scale to rate aircraft handling qualities. How does the airplane feel?

Is it sluggish? How quickly do the engines respond? Does it have a tendency to pitch up unexpectedly in response to certain control inputs? How is its lateral stability?

While most of my early flights were in the familiar T-38, I eventually had the honor of flying twelve sorties in the venerable F-4. This was an airplane with *character!* When the Edwards photographer asked me to select an aircraft for my graduation photo, I chose the F-4.

In the Test Planning phase, the students received a real-world test problem that the Air Force needed to be addressed, and the school was a cheap way to get their answers. We divided into groups of four or five, splitting up the pilots and flight engineers, and set to work. This phase tied together the skills we had previously learned. It required writing a test plan, flying the maneuvers, writing a final report, and presenting an oral report to the class.

After I left Edwards, the Air Force added a fourth section, Systems Test. Due to the expense of building new aircraft, many fewer are designed these days than in the last century. Instead, the Air Force upgrades the systems on existing aircraft. For example, the B-52 has been flying since the late 1950s. The fleet's engines, avionics, ejection seats, and even the aircraft structure have been continually modified and upgraded since the last B-52 rolled off the production line in 1962. Test pilots ensure mission readiness and safety in these programs.

Throughout our three phases of TPS, student pilots and engineers had the opportunity to fly various aircraft on loan to the school by operational squadrons. This was the most enjoyable part of the program. After assignment to an aircraft, we would take a boldface and aircraft limitations test, which required memorizing critical emergency procedures and operating limits for the aircraft. We'd study the checklist, set up a profile, then fly with an operational pilot.

My most plum flight was on the TR-1, a two-seat version of the high altitude U-2 reconnaissance airplane, like the one flown by Francis Gary Powers when he was shot down over the Soviet Union in 1960. I drove with my classmate Bob Arbach from Edwards to Beale AFB in northern California for our flight.

On our first day, we completed a high-altitude "chamber flight." The purpose was to ensure we were physiologically ready for our actual high-altitude flight in the aircraft the next day. We donned a full pressure suit (basically, a space suit) and spent about an hour in a vacuum chamber on the ground, where highly trained technicians operated vacuum pumps to set the air pressure equal to what we would experience inside the aircraft. A military airplane's *internal* air pressure is pneumatically maintained at no higher than approximately 20,000 feet, which is equivalent to the air pressure at the top of Denali Peak in Alaska. But the air *outside* the aircraft would be much thinner at 65,000 feet, which is about twice as high as where commercial airliners fly. To help us practice for the possible loss of cabin pressure in an emergency, the technicians decreased the chamber's air pressure to the equivalent of 65,000 feet altitude. A flask of water in the chamber spontaneously began to boil. At that point, it was indisputably obvious to me why I needed a pressure suit!

The next day, Bob and I each flew the TR-1 in separate flights with an operational instructor pilot. I flew performance and flying quality maneuvers while logging technical data and any unusual observations. But I most cherish the memories of the *human* experience of this flight. At 65,000 feet and above, the sky is a deep dark blue. The horizon looks fuzzy where the clouds and the ocean merge to space. Since the TR-1 is designed to fly so high and slow, it has a relatively straight wing for better lift. It was almost like being in a glider. The cockpit was eerily quiet, the silence only occasionally interrupted by the instructor pilot's voice. I could sense that the air outside was thin, and that my pressure suit—which was somewhat uncomfortable—protected my blood from boiling in case we lost cabin pressure.

At such a high altitude over the coastline of the western United States, I felt a connection to those space missions I had earlier dreamed of flying. I was viewing the Earth from far above every other airplane, while feeling the cold thin air and experiencing a lonely perspective. I believe the staff at the test pilot school was being both practical and kind by assigning me this opportunity, so I could better prepare for the future possibility of spaceflight.

Some of the other aircraft I flew at Edwards included the Navy P-3 Orion, C-130 Hercules, UV-18 Twin Otter, KC-135 tanker, A-37, LearJet-24, U-21 (the military's version of a KingAir), F-16, gliders, and even the Goodyear Blimp.

The "Qual Eval" flights each had a different goal, although we evaluated handling qualities, human factors interactions, and systems operations on every flight. As much as I enjoyed flying each vehicle, it certainly wasn't for fun. Each one of those flights required a post-flight paper or possibly an oral presentation to a school staff member, for which the student was graded.

A Very Near Miss

My most memorable flight at TPS was in the A-7 Corsair, the "A" standing for "attack," meaning close air support. The A-7 aircraft in the school's inventory were unique in that they were single-seaters. Therefore, there was no in-flight "checkout" with an instructor in the other seat. Rather, each pilot studied the flight manual indi-vidually, scheduled time to sit in the cockpit for familiarization with instruments and switches, memorized the emergency boldface pro-cedures and engine limitations, and finally took a test. After a review with a qualified instructor in the briefing room, they turned you loose in the airplane. The instructor "chased" you in a second aircraft to monitor your performance.

There were three A-7 sorties in the school's syllabus, the first being a general performance and flying qualities evaluation. The third was a checkride completing the "Systems" portion of the curriculum.

The second and most exciting flight was the optional "departure flight," or "departure from controlled flight." Due to the inherent danger of this sortie, not every pilot was permitted to fly it. A school staff pilot had to certify that you could handle it.

In a departure flight, the test pilot's goal was to intentionally stall the aircraft. The pilot puts the plane in a condition where the wings can't generate enough lift to keep the airplane aloft, so it literally drops out of the sky. To do this, you pull the throttle to idle, which is minimum power, pull up the nose of the aircraft to quickly dispel airspeed, and as soon as you feel a sink rate or a "burble" in the airframe, you stomp down with your foot on one rudder pedal. This makes the nose of the aircraft snap quickly in one direction, then drop. The aircraft then will spin like a top, although the A-7 has a more violent and unstable spin. The airframe will ratchet and snap, accelerate in torque, stop, and reverse. The closest thing I can imagine it resembles is the chaotic whipsawing of riding a bull at the rodeo.

Because you are tightly strapped into the ejection seat, you are not thrown about the cockpit, but you must nonetheless fight to hold your head still to diagnose the situation. You must be able to maintain situational awareness. Which way is the nose spinning? Are you in an upright spin or an inverted spin? Then you apply the corrective spin recovery techniques. Depending on the type of aircraft, the techniques are different, but in the A-7 the recovery procedure is to lower the nose to pick up airspeed and slowly blend in opposite rudder. You must not panic or lose awareness. The pilot must have the discipline to analyze the situation with a level head and visually stay focused on a reference point.

During a departure flight, the pilot performs this maneuver about twelve to fifteen times, including the corresponding smooth recovery. It's an exhausting and draining experience, but a pilot who is flying an airplane at the extremes of its capabilities has to know what the situation feels like and how to get out of it. The alternative is to leave a smoking crater on the floor of the dry lakebed.

During my departure flight, I had a near-midair collision with my chase pilot, Bruce Arnold. On about my tenth spin recovery, Bruce lost sight of my aircraft. During my pullout from the spin, he flew right in front of me, directly across my flight path.

He was so close I could read the writing on his helmet. Neither of us had any time to react. We missed colliding by only a millisecond.

Had we "bought the farm," I'm sure the test pilot school would have permanently stopped these departure flights. However, I needed to experience putting an aircraft into a "corner of the envelope," training myself to diagnose an abnormal situation, and keeping my decision-making skills sharp during chaos. I honed these tremendously important skills through this exercise. For a pilot—or anyone for that matter—to be a confident, experienced, and effective contributor, you need to see abnormal situations, know how to quickly diagnose what's going on, make timely decisions, and then be in a position to share what you learned with others. We cannot always be *so* safe that we sacrifice mission readiness.

Another memorable flight at the school was the T-38 "dive" flight. This mission familiarizes the student test pilot with flying a *lifting body* approach and landing. A lifting body is an aircraft with little or no wing area, so the underbody of the aircraft produces its lift—the force that allows the aircraft to stay up in the sky. Lifting bodies became popular when the Air Force and NASA began researching supersonic and hypersonic flight back in the 1960s. The T-38, although not technically a lifting body itself, had stubby wings and was a simple aircraft that was well-suited to practice these types of approaches.

The school used the T-38 for what they officially called the *low L/D flight*, where L/D is the lift-to-drag ratio. An airplane like the U-2 or a glider has a high L/D ratio, meaning that it descends quite gradually if power is cut. In an aircraft like the T-38 or the space shuttle, the L/D ratio is low, meaning that the vehicle essentially becomes a "flying brick" without the application of sufficient power to keep it aloft.

On a low L/D flight, the test pilot will climb to about 20,000 feet, set up for a steep approach to the runway, then reduce engine power so the plane sinks like a rock. These approaches simulate the characteristics of a space shuttle approach, in which the space shuttle descended about seven times steeper than a commercial jet. We flew eight of these approaches on one L/D dive flight.

My dive flight was with Colonel Steve Stowe, the school's vice commandant. This flight took place late in 1989, between my astronaut interview and the selection announcement. Colonel Stowe wasn't known for his diplomacy, and he apparently liked to intimidate and humble student test pilots.

As we taxied out for takeoff, he said, "Dick Covey is a good friend of mine." Covey was in the first class of space shuttle astronauts. He was a member of the astronaut selection board and was present at my astronaut interview, so he certainly knew who I was. Just as I was about to call the tower for permission to take off, Colonel Stowe said, "I'm going to call Covey when this flight is over, and I'm going to tell him how you did."

Oh no, I thought, *I better not mess up!*

Some people might say, "What a jerk! Why would he say such a thing at the worst possible time?" Was he deliberately trying to make me upset or nervous? Here was a teachable moment. These kinds of people will always be around, as will people who are trying to be helpful but who just seem to say the absolute wrong thing at the worst time. You can't control them. You can only control how you let yourself react to them.

My heart skipped a beat. I reminded myself, *I prepared for this flight by chair flying, mentally rehearsing these procedures over and over. I am going to do my very best.*

Since I had never flown a shuttle dive in my life, I had no idea how I would perform. But I didn't want Colonel Stowe to call the astronaut selection board and tell them I was a "ham-fist." There was too much at stake to let him rattle me.

I dropped all the emotions, said, "Yes, sir!" and locked into my high-performance mode. I stayed focused on the procedures, the altitude, the airspeed, the power settings, the configurations, and the radio calls. I made all the landings on speed and in the zone, with no help from him. It was a perfect flight, if I say so myself.

After the flight, he said, "Good job."

I'll never know if he called Colonel Covey at NASA. He certainly couldn't disqualify me now, and his attempt at adding pressure had just heightened my resolve to prove to him that I couldn't be intimidated.

Field Trip

One of the most anticipated aspects of the test pilot school was our class field trip. This happens about midway through the course, and my class departed for our trip in the second half of January 1990. Our first stop was EPNER, the French test pilot school, located south of Paris. The students and staff welcomed us warmly. We flew and evaluated the Falcon 20 aircraft, a business jet akin to the Learjet, over the stunning French countryside.

Our next stop was Italy, where we spent the day at the Italian Flight Test Center, just south of Rome. I flew a G.222, a medium-sized military transport aircraft that the United States eventually purchased and called the C-27A. A former student of mine from Vance AFB, Lt. Ammoniaci, was a test pilot at the base. It was a reversal of roles for us, and I was happy to fly with him as his student pilot. He even impressed me with a barrel roll, which only the best pilots can execute in that aircraft. We flew down the west coast of Italy, practicing flight test techniques and observing historical sites along the beautiful Italian coastline.

That night, our class decided to board the train to Rome. With only one day in Italy, there was no time to waste. Just before the train door closed, someone yelled, "Harry isn't here!" They must have needed Harry for some reason, because everyone jumped off the train just as the doors started closing. I stayed on the train with John Kirk, one of our navigators.

We looked at each other. I said, "John, the next train is in thirty minutes. I'm not waiting thirty minutes." The doors closed, and John and I went to Rome without the rest of the class.

We knew the Colosseum closed at five o'clock. We *had* to take this once-in-a-lifetime chance to see this historical treasure. We got off the train, ran over to the entrance to the Colosseum—and found it locked. The January sun sets early in Rome, and it was already growing dark.

I was terribly upset. I hadn't come all this way just to be locked out.

"John, we can get in. Let's find a place to climb the fence."

We walked to the back side of the Colosseum, away from the train station, and looked around. It was dark, and no one was in sight. John boosted me over the fence, and then he climbed over after me. We moved stealthily between the columns, darting from one to the other, pretending to be spies, and avoiding capture by the guards. There were several people inside: not guards, but janitors. What really surprised me was seeing dozens of stray cats. My imagination got the best of me. I told John we were early Christians avoiding the lions, and if a cat saw us, we were dead.

After tracing most of the inside circle of the Colosseum, we decided we'd better get out of there or else find ourselves in jail. We exited the same way we came in.

We had to wander blindly to find a restaurant. (Of course, it's not all that hard to find a restaurant in the center of Rome.) The sightseeing was great, although we completely lost track of our location.

After dinner, we learned of a random labor strike that evening by the train workers. We would have to take a taxi back to the base, but finding one was difficult due to the strike. When we finally did hail one, our lack of Italian language skills made it hard to tell the driver where we wanted to go. We had to creatively describe the location of the military base without the aid of a map. Somehow, John and I managed to get back before our flight departed for the United States.

Our class had placed a foosball table on the KC-135 to pass the time on our overseas flights, but we were all so exhausted we slept the entire way back.

Our next stop was Wright-Patterson Air Force Base near Dayton, Ohio, home of what at the time was Air Force Systems Command. I flew the T-39 Sabreliner aircraft with classmate Dave Glade. Due to the low clouds, I managed to log a real instrument approach and landing, which is something we rarely had a chance to do in the constant sunshine of Edwards.

It was here at Wright-Patterson that I first met Captain Cady Coleman, who would eventually be one of my space shuttle crewmates. I admired her clever sense of humor. She volunteered to host our class, and she pulled many pranks in the process. She assisted as well in a practical joke we played on our school commandant, Colonel ET Pollock. He wasn't able to accompany us on the trip, so we made twenty-five name tags for our flight suits, with his name on all of them. Two dozen students from Edwards Air Force Base wore ET Pollock nametags in the Wright-Patterson officers' club bar. Our behavior was only *slightly* wild. No permanent damage.

The remainder of our field trip included a visit to the Navy Test Pilot School in Patuxent River, Maryland, where I flew a KingAir and an H-58 helicopter. Finally, we stopped at the Flight Test Center at Eglin AFB, Florida. It was here that I met future astronaut Kevin Kregel.

Kevin took me up for a flight in the F-111 Aardvark. This Vietnam-era attack aircraft was capable of a variety of missions, primarily bombing, reconnaissance, and electronic warfare. It had variable-sweep (movable) wings and automatic terrain-following radar. We flew a low-level route, followed by a high-altitude supersonic flight. For the first time in my career, I exceeded Mach 2 (over 1,500 mph).

I feel fortunate to have flown this aircraft and the F-4, as both have a tremendous history of achieving advances in aeronautics and systems. Women still couldn't fly combat aircraft operationally in 1990, and my experiences in test pilot school were the only way I could get a shot at an actual flight in combat aircraft at the time.

The day after I left Wright-Patterson, Cady Coleman took her prank to the next level. She mailed a phony letter to the Flight Test Center Commander back at Edwards, Major General John Schoeppner, with a list of all the inappropriate behavior at Wright-Patterson conducted by someone wearing a flight suit with the name tag ET Pollock.

Both Colonel Pollock and I were unaware of this letter until after its delivery. Since Colonel Pollock and his wife had a great sense of humor, my class apparently wasn't too worried about potential backlash. As hilarious as the letter was, though, General Schoeppner didn't appreciate it. A few weeks later, that two-star general—four levels above me and the most senior man at Edwards—called me into his office and chewed me out over the letter. He said the wives didn't appreciate these types of jokes, as they worry about what their husbands do when they're on travel.

"I know you students like to compete to show who can pull the best pranks, but this is beyond funny. So, stop it with the jokes. Do you understand, Major Collins?"

"Yes, sir!"

Well, there goes my career, I thought. *Called on the carpet again and in trouble with my senior commander.*

Leadership Lessons

I didn't anticipate that being class leader would be among my most valuable experiences at Edwards. I didn't choose the role. I wasn't elected to it, nor was I appointed as the most qualified for the position. Military custom is that the senior ranking officer becomes the class leader by default. Since I hadn't been able to apply to the school until I finished my requirement of three years of service to the Air Force Academy, I had already competed for and achieved the promotion to major before coming to Edwards, while the others in my class were still captains.

I was a reluctant class leader. Not that I felt I couldn't do it. Rather, I wanted to focus on my flying and apply myself 100 percent to the mission.

Somewhere in the back of my mind, though, I knew that as the first woman in this position, I would be setting a precedent. Would the guys accept me, especially the pilots from the combat world, where women weren't even allowed to fly? Would my being a transport pilot from a noncombat role be a handicap? Worrying about those meaningless questions was a waste of time. I don't recall anyone from the class expressing any doubts about me as a leader. After all, I had earned the right to be here just as much as any of the other pilots.

I tried to put myself in the minds of each member of the class. What would they want in a class leader? What would they *not* want a leader to do? I began by talking to Colonel Pollock. I spoke with various instructors. I didn't have a single mentor, but I did have people whose advice I found valuable as I navigated the class leader role.

The twenty-five people in our class were the cream of the crop. All had degrees in mathematics, physics, or engineering. All had performed superbly in their previous operational flying roles. All had been recommended by their previous commanders. I felt my primary job was to assist them in getting the most they possibly could from the school experience. If our class didn't gel together well, I would feel responsible. Perhaps I was taking on too much responsibility with this type of thinking, as a leader doesn't have control over everything. But at the least, I could set a tone.

I held a class meeting on our first day together. I asked each class member to volunteer for one of the administrative positions—class treasurer, class historian, party coordinator, and many more. I knew that if I waited until classes got busy, no one would volunteer for anything. If I asked on the first day, though, people would be more willing to get involved. I asked everyone to help in at least one position. That worked well.

The other thing I learned was not to try to run the show. My role was to be there when needed. If a student had an issue, I could help solve it. If two people disagreed, I could help them work through it.

Having advanced in my career by my own decisiveness, I had to be careful not to become an autocrat. I accidentally waded into this mode at times when things got dicey or I was just plain tired. When that happened, I was reminded that this style doesn't work with highly motivated and mature individuals. They didn't want someone *leading* them. I had to adopt a much more subtle and hands-off approach, letting my classmates make their own decisions so they felt in control of their training. If no one wanted to decide about a trivial issue, then I would just resolve it myself so we could get back to our work.

I observed the capabilities and competencies in my classmates. Leaders need to know where their people's talents lie. I saw in our class a wide range of computer skills, mechanical skills, and people skills. I noticed who spoke out, who didn't care what others thought, who just followed orders quietly, who gossiped, who had the best memory, who liked creature comforts, and so forth. I realized who the natural leaders were.

As the weeks and months went by, the originality, motivation, and camaraderie of my class astounded me. We were such a tight group that we still hold class reunions every five years.

Finally, you couldn't help but feel the profound sense of history at Edwards, reminding us we were links in the unbroken chain between its early days and its ongoing role in the advancement of aviation and space exploration. When you went in the parachute room before a flight, you saw three lockers on the right-hand side: one for the Flight Test Center commander, one for the wing commander, and one labeled CHUCK YEAGER. We held our hundred-day party at the old Pancho Barnes place outside the base. It was the twentieth anniversary year of Woodstock, so we wore tie-dyed outfits!

And on January 20, 1990, we watched space shuttle *Columbia* land on Runway 22. Captain Jim Wetherbee was the shuttle's pilot that day. Five years later, he would be my commander on my first spaceflight.

Did I mention that only four days before watching *Columbia* land, NASA called to offer me a job as an astronaut?

Chapter 8

A DREAM COME TRUE

If this were not already the most momentous and challenging year of my career, another opportunity came along that I couldn't pass up: the chance to apply to the NASA astronaut program.

NASA selected astronaut classes on average every two or three years during in the 1980s, the latest one announced in June 1987. At that time, I didn't feel I possessed the qualifications of a competitive candidate.

Then NASA issued a call for applications in early 1989, shortly after my acceptance at Edwards. Talk about a dilemma! I didn't want anything to distract me from my work at the test pilot school. I also realistically knew that this might be my single best shot to apply for the astronaut program. Competition was so fierce that I might not make it on my second, third, or even a later application. It gets tougher every time, because each succeeding year, applicants learn what boxes need to be checked to become even stronger candidates.

It was now or never.

I filled out the mountain of paperwork and mustered the courage to ask the chair of the academy's math department to write a recommendation letter. He agreed, to my relief, and surprised me by saying I'd make an excellent candidate.

I was one of almost two thousand people to apply that summer—two thousand people who, like me, saw this as the chance to fulfill their lifelong dream. NASA whittled that list down to about one hundred candidates to interview.

In late September 1989, I received a call from Duane Ross, the administrative officer of the astronaut selection board. Duane had administered every selection board since the beginning of the shuttle program, and he was highly respected. He asked if I could fly to Johnson Space Center the first week of October for a weeklong interview.

I immediately agreed, although I needed permission from the school commandant to miss a week of classes. That wasn't a sure thing. I had only been at the school two months. The critical section of the curriculum dealing with longitudinal stability was scheduled that week in October. The instructor of that section, Ray Jones, was a legend in the test pilot community, so much so that the top award for flight test engineers is named after him. I didn't want to miss his weeklong class—but again, this was the chance of a lifetime.

I asked the school commandant. He deferred the decision to Lieutenant Colonel Warren Hanson, our operation officer.

Oh no! I thought. *There's no way he'll let me take a week off from school.*

Hanson was a quiet man. I was never sure what he was thinking, and he often came across as a tough personality. He seemed to be a gruff and mission-minded pilot, at least in front of the students.

As I sought the courage to ask Hanson for permission, my mind went into overdrive. I couldn't help but wonder: *Had he also applied? Was he previously interviewed by NASA, or maybe turned down by NASA? Did he feel the rest of the students would think I was unfairly getting special attention?*

I reasoned with myself, *He will start with no, so I must work on my elevator speech and be ready to convince him that this is a chance of a lifetime.*

I approached him near his office and began telling him about the call from NASA. Before I could start on my rehearsed persuasive speech, he said, "Yes."

Pilots and astronauts hate what we call a "single point failure," meaning that if something goes wrong, there is no backup or workaround. To this day, I realize that Hanson was potentially a single point of failure in the chain of events that would lead to my becoming an astronaut. Despite his reserved nature, he seemed to care about his students as individuals. He could easily have said no, and that one word might have prevented me from ever becoming an astronaut.

My task next was much easier: buying a business suit to wear to the interview. Since I had received only one week's notice, there was just one weekend to buy a civilian suit. As luck would have it, we had a school trip to Beverly Hills to attend the Society of Experimental Test Pilots convention scheduled that weekend. I skipped out of the convention for a few hours and bought a NASA-blue suit at Robinson's department store in Beverly Hills—on sale!

A Whirlwind Week in Houston

Looking back to the astronaut interview week, I'm still amazed at how at ease my mind was as I approached this examination that would decide my future. Yes, it was an exam. Every move we made was watched and judged, from the moment we met the van that picked us up from the airport until they dropped us off at the airport at the end of the week. Frankly, I was just thrilled to visit Johnson Space Center (JSC), to see the facilities that ran our space program, and to meet my astronaut heroes. I only wanted to enjoy this opportunity, I didn't want nervousness to clog my memories or disrupt my performance.

During my van ride from the airport to my hotel on NASA Road One, I met John Grunsfeld, an astrophysicist from Caltech who was interviewing for an astronaut role. John didn't make the cut that year, but NASA selected him during the next round in 1992. Years later, John would be the astronaut guru behind the last three of the Hubble Space Telescope repair missions. He would eventually become NASA's chief scientist.

I learned that ours was the second group of interviewees, out of five groups total. I was one of this week's twenty candidates. NASA would interview about one hundred people this fall and select a yet-to-be-announced number of pilots and mission specialists. The pilots would train for flying the shuttle on launch, orbit, entry, and landing and could eventually command space shuttle missions. The mission specialists would perform spacewalks, operate the shuttle's robot arm, and conduct science experiments.

Our first official interview day was a Sunday. That morning, we took written psychological exams. As I met the other candidates prior to sitting down, I could see how diverse we were in background, experience, and our mental states. Some of us were talkative and friendly, some quiet, and others clearly nervous. We spent a long day taking timed written tests like the Minnesota Multiphasic Personality Inventory. After about five hours, we were fed up with answering questions like "Do you love your mother?" What a relief to go back to the hotel for dinner and conversation!

We spent about 90 percent of our time over the next five days in medical screening. This included the normal expected exams like vision, hearing, and doctor's evaluation, and also more invasive procedures—mammograms for the women and proctology for everyone. We were wired up with heart sensors (the men had their chest hairs shaved, then the skin cleaned with alcohol, *ouch!*) and ran on a treadmill. The doctors wanted to see our heart function at a specific high rate, and if we didn't reach that rate, we had to run harder.

I was only thirty-two years old, and I wondered why I had to go through these procedures. The answer was "NASA will be investing a lot in you. We can't have you removed from flying status at a young age for some physiological problem we could have foreseen. We need you to be healthy for decades to come!"

There were more unusual tests. My favorite was the claustrophobia screening. A technician attached a blood pressure cuff to my arm, sweat sensors to my fingers, and those pesky heart monitor stickers to my chest. I crawled into an inflated balloon, about three feet in diameter, which looked something like today's exercise balls. Once I was inside and curled into the fetal position, the technician zipped up the back of the ball. For obvious safety reasons, they hooked up a hose to keep air circulating. Examiners told me I would pass this test if I stayed inside for thirty minutes. That was a piece of cake. I fell asleep, just as I had done in the Air Force POW camp isolation box twelve years earlier. I was so relaxed that I actually felt sad to come out—like having to get out of bed on a Saturday morning.

I pondered the people I knew who were seriously claustro-phobic. I asked my technician, "How many people have failed this test?" She said only one person in the history of astronaut candi-dates had failed. Perhaps those who are claustrophobic don't even bother to apply.

There were several more enjoyable events of the interview week. We were each assigned to watch a simulation in Mission Control. We went to Ellington Field to view the T-38 aircraft and then visited the Weightless Environmental Training Facility, a huge pool used for spacewalk training.

Thursday night we had a social event at Pe-Te's Cajun BBQ near the airfield. On the bus ride over to Pe-Te's, some of the candidates said, "This event is the *real* interview. They want to know if you are socially inept. Don't say anything stupid!"

I still wonder why some of our candidates put so much pressure on themselves. I looked to the social event as a chance to meet more astronauts and hear about their individual experiences. I had a great discussion with Mike Lounge, an astronaut who flew on the first mission after the *Challenger* accident. He accused me of "doing my homework," and I laughed. I had closely followed his mission just one year earlier, as his crew so bravely led the Space Shuttle Program back to successfully flying again. Little did I know then that I would eventually *command* NASA's next return-to-flight mission, sixteen years later.

Our hosts encouraged us to walk over to Building 4 in our free time to visit the current astronauts whose offices were on the third floor. I went over a few times but unfortunately found only a couple astronauts, as most were out working somewhere else.

The most feared part of the week—even more than the actual interview—was the hour with the psychologist. I was actually looking forward to having a good time talking with this guy. Many of my fellow candidates were petrified.

After about thirty minutes with the shrink, I started to become concerned. First, he told me I got seven questions "wrong" on Sunday's personality and psychological exams.

He asked me, "Tell me about your hallucinations."

That floored me. "Huh? I don't have hallucinations."

"But you answered 'true' to the question, 'When I walk down the street, I see things other people don't see.'"

"Yes, I did answer 'true.' When I am with my husband, for example, I'll notice flowers, landscaping, and people's clothing, while he has no clue they're there. He looks at other things."

"Well, that question is designed to find out if you hallucinate."

I started feeling defensive. "Sir, with all due respect, someone who wants to be an astronaut is not going to confess to unusual mental issues."

The interview went on and on like that. Then he started asking me about my mother. "Do you love her?"

"Good grief, of course I do!"

Then he asked me, "Are you doing this for *her*?"

I lost it. "No, I am not doing it for her! She doesn't want me flying jet airplanes around the world, risking my life as a test pilot, and now wanting to strap myself onto a rocket, just a few years after the *Challenger* accident!"

I am not sure why he didn't disqualify me at that point. Someone later said that the astronaut selection board doesn't actually listen to the shrinks unless there is some kind of serious concern, so I shouldn't worry about that part of the interview.

I faced two risks of "DQ" (disqualification). The first was the hearing test. The technician put the headset on me, but she didn't place it precisely over my right ear. It was a simple test that I easily passed every year in the Air Force. I listened to the tones, pushed the little button as requested, despite the noise in the hallway.

Here was an important lesson. *You* are in charge of ensuring proper collection of your data. I should have either correctly placed the headset or asked her to fix it before we began the test. Instead, feeling confident, I thought: *No big deal. I have never failed a hearing test.*

Well, I failed this one.

Later that day, Dr. Richard Jennings, the chief doctor of the clinic, called me in and asked in a loud voice, as if I were deaf, "What is wrong with your hearing?" I explained what I thought was the problem, and he sent me back for a retest. This time I put the headset on myself, and I asked the technician to close the door to the hallway so I didn't have to listen to extraneous noise. Result: I passed!

The second and more serious risk of DQ was the eye exam. I had issues with passing the visual acuity part of the Air Force test when I was in college and my first year in pilot training. The acuity requirements were looser for an astronaut, though. Instead of 20/20 for both near and far vision, a pilot could be 20/40 uncorrected but had to be 20/20 with glasses. I passed this part with no problem. However, I failed the primary depth perception portion of the exam. I later learned that visual depth perception is the number-one reason pilot candidates are DQ'ed. The doctor let me take an alternative depth perception test with a different kind of target, and I passed it easily.

Starting with the next class selection in 1992, NASA eliminated the alternative test, and all pilot applicants had to pass the primary test or be DQ'ed. That was the new standard practice until the year 2000. That year, when I went in for my annual medical exam, the eye doctor said to me, "We're changing our depth perception eye test to now allow the alternative test again. The medical board figured that since *you* successfully landed the space shuttle with no problems, maybe our requirement was too strict."

If I hadn't been selected as an astronaut in 1990 and had to reapply in 1992 or later, *I could never have been a space shuttle pilot*, because I couldn't pass that primary depth perception test. Another lucky break.

The Interview

The crux of the week was the Big One—my in-person interview with the selection committee. This board had about a dozen people, chaired by former flight director Don Puddy, the chief of the Flight Crew Operations Branch. A candidate who interviewed before me said that Mr. Puddy fell asleep during his interview. I suppose twenty interviews in a week can get pretty boring for the selection board.

I was lucky to have some free time before my interview. I went back to my hotel room, exercised for about half an hour, showered, ate, and took the bus to the Astronaut Crew Quarters. It was a simple one-story building back in the northern corner of Johnson Space Center. I walked in, said hello to administrative assistant Teresa Gomez, and took my seat in the lobby to wait my turn.

I was excited but not particularly nervous. I was looking forward to this chance to meet the astronauts and also to talk about my qualifications and how I could benefit the space program. Since I was still a student at the test pilot school, I thought they probably wouldn't be serious about hiring me this time around, which lifted much of my performance pressure. Some people had suggested that maybe the board just wanted to check me out for consideration in future years.

During the shuttle program, prospective astronauts applied either as pilots or as mission specialists. The military would forward your application as a pilot only if you had graduated from test pilot training, either the Air Force's school at Edwards or the Navy's program at Patuxent River. Since I hadn't yet *graduated* from test pilot school, the Air Force followed its rules to the letter and had submitted my application as a mission specialist rather than a pilot. I was happy just to be able to interview.

The pilot who interviewed before me walked out of the meeting room with a stone-cold look on his face. "Hi!" I said. He just shook his head and walked past me.

Hmmm, I thought. *This may not be pleasant. But I will have a positive attitude, and I will not put them to sleep!*

I walked in and shook hands with everyone. I sat near the top of the open-ended rectangle of tables and waited for the first question. John Young, Charlie Bolden, Hoot Gibson, Dick Covey, Mike Coats, Rhea Seddon, Jerry Ross, and Mary Cleave were some of the questioners.

I flubbed the very first question. Jerry Ross asked me, "Let's start with high school. What did you do in high school?"

My mind started racing. As you know, my high school days were not the best. I was just sort of *there*. I wasn't an athlete, I wasn't a math whiz, I didn't have the lead role in school plays. I mostly worked part-time jobs—and in the moment, I forgot about my part-time jobs. I drew a blank when I tried to think of something meaningful to say.

I started chattering like a contestant in a beauty pageant. "I wish I had applied myself more in high school. There are so many great opportunities. My message to young people today is to take advantage of the great opportunities. There are so many. Take a risk. If you fail in high school, who cares? You will be preparing yourself for the future. Take lots of courses, play a sport, don't quit five minutes before you sing for the lead part, make friends, do some volunteer work, talk to people, don't be so shy, you can do it." Basically, I was thinking about all the mistakes I made in high school.

After an awkward moment of silence, Jerry looked me right in the eye and said, "Oh come on, weren't you a cheerleader?"

What a scene. Where did *that* come from? I recalled trying out for cheerleader three times in high school and failing to make the cut each time. I did cheer for my church league in ninth grade, though.

I finally answered, "Only one year." It was an awkward minute.

Fortunately, after that things got much better. I learned a lesson here. I had started out on the wrong foot because of a question about a period of my life that I wasn't particularly proud of, much less able to answer about with some substantial evidence. I recovered, though, and I didn't let that misstep throw me off balance for the rest of the hour. In fact, I enjoyed the interview. I felt I handled the remaining questions well. I drew a few smiles here and there, especially when the interviewers went on to joke with one another.

"Why do you want to be an astronaut?"

"What does your husband think of what you are doing?"

"What kind of car do you drive?"

The legendary John Young—Apollo 16 moonwalker and commander of the space shuttle's maiden flight—asked me, "Have you ever been afraid in an airplane?" He knew, as well as I did, that *every* pilot has been afraid in an airplane.

Instead of simply answering, "Yes," I began, "Which time would you like to hear about?"

Dick Covey asked me, "What do you think of the F-4?" I knew John Young was the first pilot to land an F-4 on an aircraft carrier, many years earlier during his Navy career. Despite all the test pilot complaints about the terrible flying qualities of the F-4, I had to say something positive.

"Well . . . It's an airplane with lots of *character*," I honestly answered.

Early in the interview, somebody asked me, "Wouldn't you rather be a pilot?" I had anticipated that question and replied, "I applied as a mission specialist and as a pilot. I would be happy to be either one."

It seemed like everyone else on the panel asked me the same question. Even the last interviewer said, "Come on, you'd really rather be a pilot, wouldn't you?"

I said, "I'm going to stick to my first answer. I'll do either one."

I actually felt that way. All I wanted was to be an astronaut.

When I walked out of the room, I felt light on my feet. *What a great group of folks*, I thought. *A group of people I would love to work with.*

I had dinner with one of the other applicants after our interviews. We compared notes. He asked me how I had answered several questions, and I told him. He said, "No, that's not what they want to hear. You should have said . . ." I didn't let that rattle me.

And no, he didn't make the cut.

When I returned to Edwards after my whirlwind trip to Houston, I tried to put the interview out of my mind and concentrate on school again. I needed to make up for the missed week of training.

But I couldn't stop wondering: *Will they select me? And, if so, will it be as a pilot or a mission specialist?*

The Call

January 16, 1990, began with a Qualities Evaluation flight in an A–37 light attack aircraft. After returning from the flight, I walked past the school's administrative office, and I glanced at the message board. I saw my name on a single yellow message paper tacked to the bulletin board. It read: "Major Collins, call Duane Ross at NASA."

My heart started beating a mile a minute.

This is it. My future. My dream. The fulfillment of my work throughout my professional life.

Cool down, I thought. *Maintain an even keel.*

I walked over to the landline and dialed the number. The sergeant in the admin office watched me. She knew what was going on and shared in my excitement. Teresa Gomez answered the phone at Johnson Space Center.

"Hello, this is Eileen Collins returning a phone call from Duane Ross."

"Oh yes, just a minute."

Duane got on the phone. "Hello Eileen, I'm going to connect you to Don Puddy."

Mr. Puddy was not an astronaut. The rumor was, if you were *not* selected, an astronaut would give you the bad news. This was promising!

He got on the phone and said: "Major Collins, I have a guy here who really wants to talk with you."

Hmmm, I thought, and said, "OK."

The next thing I heard was a voice with a deep southern twang: "This is John Young. Do you still want to come to Houston and work for us?"

"Yes, sir!"

John went into a long list of all the things that Johnson Space Center does, spending perhaps five minutes describing Mission Control, space shuttle engineering, T-38 flying, simulators, studying future travel to the Moon and Mars, and the like. I can hardly remember any of what he said.

He concluded with, "Do you have any questions?"

"Yes, sir. Will I be a pilot or mission specialist?"

He laughed. "Pilot! Yes. Pilot. You will be the first woman to pilot the space shuttle!"

Reflecting on the Achievement

Long after the interview, I learned about some of the things that went on behind the scenes with the selection board. NASA knew that the first woman selected as a pilot would be under intense scrutiny. Her reputation—and NASA's—would be on the line. NASA had previously interviewed two other female pilots, but the selection board had reservations about them.

Astronaut Mike Coats later said, "It's not fair and it's not right, but the first person to break a barrier is under tremendous pressure, and it has to be somebody that almost can't fail. . . . We needed the first female pilot to be somebody who was a slam dunk."

Astronaut Mary Cleave was apparently the first on the selection board to recommend considering me as a pilot. The board extensively investigated my performance throughout my career. Mike Coats joked that they even interviewed my kindergarten teacher! I'm happy that everyone they talked to vouched for my abilities and my character.

With as much humility as possible, I felt I was up to the challenge and I wouldn't disappoint them. I felt a sense of mission and a renewed dedication to the importance of space exploration. My new profession would incorporate my lifetime of curiosity, scientific interest, and love of problem solving. And I would not only be the first woman to pilot the space shuttle; I would be opening up opportunities for other women to fly in space. I knew I had to be the best I could be.

If I performed well, other women would have these opportunities, and young girls could aspire to these possibilities.

In hindsight, piloting was the correct role for me. That's where my experience and expertise lay. Although I would have enjoyed being a mission specialist, I had been flying since I was twenty years old, and that was my talent and passion. The agency was ready to select a woman pilot, and I'm honored that they felt I was the right person at the right time.

After receiving the call, I felt more relieved than excited. I didn't feel like celebrating or buying drinks for everyone at the Officers Club or calling everyone I knew. Instead, I experienced a profound sense of calm, of peace. A huge weight had lifted off my shoulders. It just felt *right*. There were many places the Air Force could have sent me after test pilot school, and I probably could have been happy with any of them. I felt called to fly the space shuttle, though, and I knew it was where I belonged.

I had to keep the news secret from everyone except Pat until NASA made the official announcement the next day. There were several other pilots at Edwards who had interviewed but were not selected. I worried about how they would take the news about me, especially since I hadn't even graduated from the program yet. If anyone resented me, though, no one ever said anything.

My parents and siblings were stunned at the news. Although I had dreamed of this day since I was a little girl, it was a dream I kept entirely to myself for more than twenty years. Pat knew, obviously, but my family back home had no idea that I was even *considering* a career in the space program.

Few of my colleagues in the Air Force knew, either. Frankly, I rarely told anyone I wanted to be an astronaut because I suspected they'd tell me that I couldn't do it.

My brother Jay said that he and my mom just stared at each other in disbelief when I told them the news over the phone. My sister, Margy, witnessed *Challenger* explode from her home in Orlando, and of course she was worried about me. I reassured Margy that I was confident NASA had found the problems and corrected them. I had no reservations about flying on the space shuttle.

I still had half a year of test pilot school ahead of me before I was to report for duty at Johnson Space Center. My astronaut position was contingent upon my successful completion of the program.

I graduated in June 1990 and reported for duty in Houston in July. For once in my career, after years of frustration, the timing worked out perfectly!

The Collins family at their farm in Ridgebury, PA, circa 1912. My great-grandparents, James and Nora Collins, are seated; and my grandfather, Francis Collins, is second from left. The other four are my great-aunts and great-uncles: Nonie, Jeremiah, Margaret, William, and Nellie Collins. My Collins ancestors came to America from County Cork, Ireland, during the potato blight, about 1850.

My mother, Rose Marie Collins, with Edward, Margaret, and me at home in Elmira, NY, in 1960.

Kindergarten graduation, 1961. I attended Hoffman School (now closed) in Elmira.

The Collins children (I'm in the foreground) at Harris Hill, Elmira, in the spring of 1969. Harris Hill was the "Soaring Capital of the World" at that time. My dad, Jim Collins, brought us here regularly to watch the gliders take off and land. The National Soaring Museum is at this location today.

My mother stands on the porch of our family's home, near the intersection of West Second and Columbia Streets in Elmira, as the flood waters from Hurricane Agnes rose around us on June 23, 1972. We were evacuated shortly after this photo was taken. The flood devastated the city.

An Air Force ROTC staff meeting, Syracuse University, 1978. The cadets held leadership positions and ran the corps by executing military formations, conducting teamwork exercises, and holding community service events.

With a T-41 trainer, the Air Force equivalent of the Cessna 172, during Flight Screening Program ("Fishpot") in August 1978. FSP screened student pilots for basic flying competency and cleared them for the more difficult Undergraduate Flight Training Program.

Heading out to fly as a T-38 student at Vance AFB, Oklahoma, in 1979.
I decorated my helmet with my initials, which happen to be the key
terms of Einstein's famous $E=MC^2$ equation.

Graduation from flight school at Vance AFB in August 1979,
with my beloved T-38. (USAF)

On December 4, 1980, I became the first woman pilot to fly in an F-15. I'm with instructor Capt. Yates in front of our aircraft at Holloman AFB, New Mexico.

With my Danish Air Force student following a training sortie in May 1981. Note the "bag" in the student's cockpit, which could be pulled forward during flight to block the view through the canopy and simulate instrument conditions of flying through clouds.

Check Section pilots at Vance AFB in 1982. Check Section was composed of the most feared instructor pilots—those who gave check rides to student pilots before they could proceed to the next training phase. I was the only woman pilot in Check Section, as well as the first and only woman instructor pilot at Vance during my three-year tour.

Our C-141 crew during a European mission in November 1983. Crews typically included a commander, a copilot, two flight engineers, and at least one loadmaster.

Pat Youngs and I at home after our August 1987 wedding.
I was an assistant professor of mathematics, and Pat was a golf
coach and athletic instructor at the US Air Force Academy.

My graduation photo with one of my favorite aircraft, an F-4 that
I flew throughout the program at the Air Force Test Pilot School
in June 1990. I left Edwards AFB immediately after graduation to
report to NASA as an astronaut candidate. (USAF)

NASA's Astronaut Group 13, "the Hairballs," during training at Vance AFB in August 1990. Left to right, from top: Dave Wolf, Rick Searfoss, Ken "Taco" Cockrell, Charlie Precourt, Michael "Rich" Clifford, Don Thomas, Bill McArthur, Bernard Harris, Bill "Borneo" Gregory, Jim Newman, Susan Helms, Ellen Ochoa, Nancy Sherlock, Dan Bursch, Tom Jones, Peter "Jeff" Wisoff, Carl Walz, Leroy Chiao, Ron Sega, me, Janice Voss, Terry Wilcutt, Jim Halsell. (Courtesy Tom Jones)

Astronauts spend many hours flying in T-38s to keep their minds and bodies conditioned for the rigors of space flight. (NASA)

The Shuttle Training Aircraft (STA) during a simulated shuttle landing approach at White Sands, New Mexico. This steep dive was the same glide slope as a returning space shuttle after dropping below the speed of sound. Astronauts had to fly 1,000 approaches in the STA to qualify to command a space shuttle mission. (NASA)

With Pat after the launch of STS-52 in October 1992. As a Cape Crusader, I was escorting some of the STS-52 crew's extended families at KSC's Banana River viewing area. On my right sleeve is the Astronaut Group 13 patch.

With the closeout crew after the successful launch of STS-55, the German Spacelab mission. As the astronaut support person on the closeout crew, I assisted with strapping in the flight crew and securing the White Room at the launch pad. We had to extract the crew from the shuttle after an earlier aborted launch attempt in March 1993.

The crew patch for STS-63 depicts *Discovery* and Mir flying in formation. The SPACEHAB module and SPARTAN satellite can be seen in the shuttle's payload bay. The six rays of the rising sun and the three stars symbolize our mission number. (NASA)

Seven of the surviving members of the Mercury 13 came to KSC as my special guests to watch the launch of STS-63 in February 1995. From left to right: Gene Nora Jessen, Wally Funk, Jerrie Cobb, Jerri Truhill, Sarah Ratley, Myrtle Cagle, and Bernice Steadman. (NASA)

Discovery lifts off at 12:22 a.m. Eastern Time on February 3, 1995, with a woman in the pilot seat of an American spacecraft for the first time. (NASA)

Our crew in the aft of *Discovery*'s flight deck during our mission.
Top row, left to right: Vladimir Titov, Mike Foale, and Janice Voss. Front row: me,
Jim Wetherbee, and Bernard Harris. (NASA)

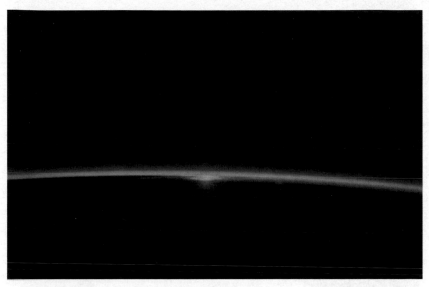

The rising sun peeks through the thin layer of atmosphere that separates
Earth from the void of outer space. (NASA)

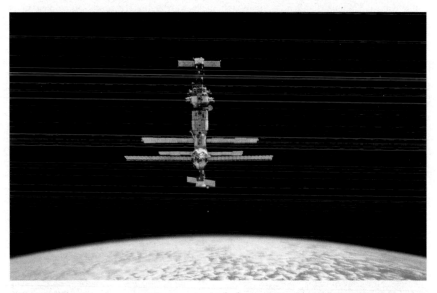

Mir's solar panels and radiators reminded us of the wings of a dragonfly as
we approached for our historic rendezvous on February 6, 1995. We were the
first Americans to see a Russian space station up close. (NASA)

Cosmonaut Valeri Polyakov watches through a large porthole as *Discovery* draws near. His fourteen-month stay on Mir is still a space endurance record. (NASA)

I'm running a checklist to troubleshoot *Discovery*'s leaky thrusters. The bag of cereal squares attached to my seat was about all I had the appetite to eat early in the mission. (NASA)

Volodya Titov and I try to deal with the endless printouts of time line changes sent up from Mission Control during our mission. (NASA)

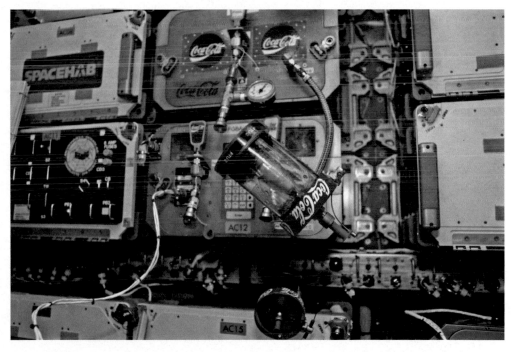

The Fluids Generic Bioprocessing Apparatus (FGBA), better known as the "Coke Experiment." This device on SPACEHAB mixed the syrup and carbonated water into a clear cylinder for us to taste and evaluate. (NASA)

The STS-84 crew patch. The six stars surrounding Mir (in Cyrillic letters) symbolize that ours was the sixth mission to dock with Mir. The Greek letter *phi* with one star symbolizes Phase One of the International Space Station program. (NASA)

A fisheye view of Mir taken through an overhead window on *Atlantis*'s flight deck. At left is the Spektr module, which was rammed by a Progress supply ship a month after we left. Mir's solar arrays show the effects of long-term exposure to the harsh environment of space. (NASA)

Chapter 9

BECOMING AN ASTRONAUT

Pat and I moved to Houston right after graduation from test pilot school. I reported for duty on July 16, 1990, one of the twenty-three members of NASA Group 13. I was now officially an Astronaut Candidate, or ASCAN for short—pronounced "ass can."

Susan Helms and I drove to JSC together that morning. We were friends from our days in graduate school at Stanford as well as our years teaching at the Air Force Academy. Susan had encouraged me to apply to the test pilot school while I was in my first year of teaching. She applied to Edwards and attended as an engineer two years before me. Now we were both ASCANs.

I felt energized as I prepared to meet my fellow future astronauts in the same room where we had interviewed last October. I had an embarrassing faux pas while getting a drink before sitting down. I walked up to Dave Wolf, a medical doctor with an undergraduate degree in electrical engineering, and I introduced myself. "You already introduced yourself, over there!" said Dave. We just broke out laughing. "Well, here we are!" We were all smiles.

Don Puddy sat us down and said, "Welcome to Johnson Space Center." We were ready for his congratulations and all the niceties a new commander typically likes to gush. But no—not today. Mr. Puddy said, "You are coming to NASA at a difficult time." He showed us a clip from the previous Sunday night's *60 Minutes* program, which asked, "What's wrong with NASA?"

This year was a particularly tough and embarrassing one for NASA. After a long delay following the 1986 *Challenger* accident, NASA deployed the Hubble Space Telescope in April 1990, only to find out that its mirror was fundamentally flawed, and its $1.5 billion mission was in jeopardy. NASA removed two astronaut commanders from flight status for a year because of unapproved activities in airplanes. The entire space shuttle fleet was indefinitely grounded due to fuel leaks recently discovered on *Columbia* and *Atlantis*. NASA had to investigate and correct those leaks before any shuttles could fly again—and it was still less than two years since the "return to flight" after *Challenger*.

NASA's reputation—and perhaps its very existence—were on the line.

"NASA has a public perception problem now," Mr. Puddy continued. He emphasized to us that we needed to understand this and operate in an appropriate manner.

Well, that put a slight damper on our day—but, I want to emphasize, it was only slight. We were all elated to be there, and we were all ready to start a year of training, of growing closer to our dreams of flying in space.

We headed over to Building 9, home of the space shuttle mockup trainers, to meet the press. Our group included seven pilots and sixteen mission specialists. We each introduced ourselves with enthusiasm, expressing confidence in humankind's future in space and conveying our desire to make a meaningful contribution to that effort.

The press didn't want to hear about our lofty dreams. Instead, they focused their attention on those of us who were a novelty in their eyes, primarily the five women in our group. Nancy Currie was an Army helicopter pilot and had been a NASA engineer. The brilliant engineer Janice Voss had a PhD in aeronautics and astronautics from MIT. Mission specialist Susan Helms would eventually become the first American military woman to fly in space. Ellen Ochoa was the first Hispanic woman selected as an astronaut; she would eventually become the director of Johnson Space Center. And there was me, the first woman pilot in the history of the US space program. My classmate Tom Jones said that he and the other guys were happy to let us women deal with the reporters!

Getting Started

Among our class's first tasks was to name ourselves and design a patch. The patch design took on a life of its own over the next several weeks. Since we were the thirteenth group of astronauts, we designed a patch with a black cat and the number 13, and named our class "the Hairballs."

Our mentor astronaut, Kathy Sullivan, told us that NASA Headquarters would never approve our patch. The shuttle mission that was originally to be STS-13 (before NASA renumbered the missions) submitted to headquarters a patch with a black cat on it. Management rejected it. "Don't bother," Kathy said, "it's already been tried." I don't remember if we actually sent it to anyone in the chain of command.

Some other class members designed a patch with the Earth, Moon, and Mars. The NASA PR people were already describing our group as the "space station astronauts" and adding that some of us would go to the Moon someday. (How did *that* work out?) We christened a more conservative and traditional patch. But the Hairballs name still lives on today.

Later that week, while we were still wide-eyed and learning our way around, the chief of the Astronaut Office, Dan Brandenstein, called a meeting for our class only. This was one of the most memorable meetings from my years in training. Dan lost several of his classmates and friends in the *Challenger* accident just four years earlier. We greatly respected him from the outset, and we were thankful for his advice.

He talked to us about fairness, a subject I had never heard a boss discuss before. He said, "Nothing here is fair. I cannot make it fair. One of you will fly first, and one of you will fly last. It's not fair.

I can't fly all of you at the same time. Therefore, I want you to be thankful you're here, and I want you to all support one another. When the first of you is assigned to a flight, the rest of you should do all you can to support him or her and that flight. Eventually, you will fly, and that person will be supporting you."

The astronaut selection board looks for each candidate's ability to get along with others, work together, communicate effectively, and keep a positive attitude even when things are falling apart. That's a tough trait to spot and grade. The selection board does the best they can. Sage advice from the boss, delivered without preaching, can prevent a lot of problems and stress by setting the right tone. Dan's words helped our class encourage one another through all the tough times we faced during the ensuing years.

Flying the NASA T-38s

As a pilot, flying was my first priority in ASCAN training. NASA operates T-38 aircraft as spaceflight readiness trainers, the same as my beloved Air Force T-38s. Astronauts fly the T-38 in the *real world*, not the structured and safe environment of a simulator. Astronauts need the simulators for shuttle mission training, but there is no better way to prepare for stressful situations than flying in an actual high-speed airplane.

If you crash in a simulator, you reset the computer and try again. If you crash in a T-38, you are done for!

T-38 pilots maintain and hone their stick-and-rudder practices, hand-eye coordination, flight planning skills, and real-time decision making, all while monitoring checklists and talking on the radio with air traffic controllers. Pilots and mission specialists practice aerobatic maneuvers and maintain physical fitness under high-g, zero-g, and negative-g stresses—trying to stay comfortable breathing oxygen through a mask and wearing a helmet, gloves, and anti-g suit, and all while strapped into an ejection seat. The T-38 was a relatively inexpensive way for the crew members to increase their confidence and keep their decision-making skills sharp in a fluid environment with real-life consequences.

What do I mean by "real-life consequences"? On October 17, 1990, I was flying a T-38 out of Ellington Field with my classmate Tom Jones in the back seat. Tom had been a B-52 pilot, so he had plenty of experience flying airplanes, but he was an astronaut mission specialist. NASA rules permitted only shuttle pilots or commanders to fly the front seat of the NASA T-38. Although Tom and I shared stick time on this flight, I was in command.

We had been flying acrobatics out over the Gulf of Mexico. It was a beautiful day. No problems. A great flight. Tom was flying, and I was working the radios. Returning to Ellington, with just enough gas to fly one practice approach and then land, we listened to the ATIS (pilot weather reports). A cold front was approaching Houston from the north. Tom and I had asked for vectors to Runway 17, which would have us fly north of the airfield and then turn and make a long approach to land headed south.

As we drew closer to Ellington Field, we could see an ominous line of thunderstorms approaching from the north. The ATIS information was slightly old. I was getting a bad feeling. There were many airplanes in the area, and we could hear them on the Approach Control frequency, asking for vectors around the storm.

We needed to land immediately. I asked Approach Control to cancel our practice approach and give us a vector directly to the field.

"Stand by," they answered. The heavy traffic and the rapidly changing weather clearly had overwhelmed them.

I could see the field directly in front of us, maybe two miles away. We didn't have enough gas to go anywhere else. I remembered when Apollo astronaut Pete Conrad was in a similar situation—out of gas and diverted to Bergstrom AFB in Austin because of the weather. He ended up having to eject from a perfectly good airplane because he didn't have enough fuel to make the runway. I wasn't going to allow that to happen to us!

The heavy dark line of rain was almost on us, and lightning bolts hit the ground in the vicinity. I was about to lose sight of the runway. I knew I needed to declare an emergency.

Recalling my earlier encounter with a hailstorm in a T-38, this was the moment I needed to take aggressive action and use all the tools I had available. I asked to proceed directly to the airfield for an immediate landing.

No answer from Approach Control. The winds were now gusting to thirty-five knots. No time to wait for permission.

"I am switching to tower frequency."

"Approved," said the controller.

I took the stick from Tom and said, "I have the aircraft," the mandatory call for handover of flight control. I pulled back the power, slowed from 300 to 240 knots, aggressively turned left, pulled off the airspeed, and dropped the landing gear and flaps. I set up for a short right turn to Runway 35, headed into the wind.

"NASA 924 is right base, gear down, full stop."

"Cleared to land," they replied.

We touched down. Within seconds, heavy rain and high winds hammered us. I pulled off the runway and waited for the worst to pass. When I taxied back to the hangar in the heavy rain, the crew chiefs opened the hanger doors and let me taxi in with engines running. In all my years of flying for NASA, that was the only time I was ever permitted do that.

Another T-38, piloted by shuttle commander Ron Grabe, had been right behind us. He had to divert to Galveston and land on a runway not generally approved for T-38s, as it was too short for normal landings.

Why do I tell this story? Whether you're a pilot or not, one should be trained for unusual, unpredictable, and emergency events. There is no time to develop a strategy when you're in life-threatening circumstances. Years of chair flying and reviewing potentially bad situations came to use that day. If you're well prepared, the adrenaline surge allows the pilot's brain to operate at high speed, so decisions can come quickly and correctly. When lives are in danger, no one wants a pilot who panics.

I cannot emphasize how much preparation came into play that day. I didn't have to hesitate or wonder what someone would think of my decision to turn off Approach Control and go visually directly to tower. We had no seconds to spare. It could have ended up with us being struck by lightning, diverted without enough gas to make it to an emergency airfield, or as a smoking hole off Runway 35 at Ellington.

This time, Tom and I survived to brag about it and fly another day.

Class Training

We spent long hours in the classroom. The vast majority of our time concentrated on space shuttle systems. All astronauts had to learn about the main engines; the solid rocket boosters; the external tank; the hydraulic, electrical, environmental, propulsion, navigation, communications, and computer systems; and many more. Each system had a paper workbook that we had to read before the associated class started. There were three briefings for each system, one each given by an engineer, an operator like a flight controller, and an astronaut trainer.

After the classroom sessions, we entered a low-fidelity simulator called an SST, or Single System Trainer. These were nearby and easily accessible if an ASCAN wanted to spend extra time studying a given system.

There were no tests. This was the "big boy program." We were all keenly aware that our lives depended on what we knew. We each felt motivated to learn as much as possible and to be able to apply that knowledge in the real world. We were mature enough to know how we could best learn and make use of this critical information.

Enrichment sessions supplemented our core curriculum. These topics included Earth observation, astrophysics, living in space, the history of NASA, discussions by retired astronauts and flight directors, and visits to every NASA Center.

We endured a shortened version of the land survival training program at Fairchild AFB in Washington, where I had completed the longer course twelve years earlier—thankfully, minus the POW camp! We attended the Navy's water survival training program in Pensacola, and we took the parachute/parasail training program at my old stomping ground at Vance AFB in Oklahoma. I vividly remembered my time as a student there, when the first shuttle astronauts visited the base and inspired me to apply myself.

Finally, every ASCAN had to become SCUBA certified. Toward the end of our first year, we completed this long program at the JSC pool and in Galveston Bay.

On January 16, 1991, our classmate Bill McArthur hosted a party at his home, celebrating the one-year anniversary of the day we all received "the call." That night, the United States invaded Kuwait to drive Saddam Hussein's army back into Iraq. We had been following the news since August, when the United States and its allies began Operation Desert Shield, the buildup of coalition forces in the Middle East. That evening, our class watched bombs go off as the war played out on national TV. Many of my fighter/bomber pilot classmates would have been in the invasion—Operation Desert Storm—had they not been astronauts. Even I might have had the opportunity to support the war as a C-141 pilot. It was an historic night, and I could sense many of my classmates felt that they should have been part of it.

By the time April rolled around, we were all clearly restless. We still had several months of training left in the ASCAN curriculum. We were tying up loose ends in the systems training and finishing our trips to the various NASA Centers.

On one trip to Stennis Space Center in Mississippi, a few men in our group started getting slightly silly and began playing practical jokes. The five women in our class decided this was just a bit juvenile. The guys were no match for Ellen Ochoa's rapid-fire comebacks. We women named the guys The BaBas, meaning the Bad Attitude Boy Astronauts. The guys fired back by naming us The WaWas, the

Whining Association of Women Astronauts. After nine months of memorizing the endless NASA acronyms, we were part of the culture and coming up with our own!

We couldn't help but bond into a tight-knit group after spending so much time together. We actually got along wonderfully as a class. We kept a scrapbook called the "Hog Log," in which we recorded our dumb jokes or pranks and memorable quotes from our classmates.

From my earliest days in pilot training, and even at the start of the ASCAN program, people thought it would be funny to nickname me "Crash Collins." That name never stuck—not after they saw me in action. Somewhere along the line, I earned the call sign "Mom." I enjoyed being in on the jokes and pranks, but I felt an instinctive need to keep us all focused and draw the line between fun and our serious work. I always seemed to be the one reminding the class when we needed to be somewhere at a certain time. "Yes, Mom!" was the habitual reply. They also nicknamed me Mrs. Cleaver (from *Leave It to Beaver*).

NASA's managers could tell we were ready and eager to move on with our careers. Rather than make us wait until our one-year anniversary in July, our leaders assigned us to operational jobs at the nine-month point. Now we would truly become important contributors to the space program.

Astronaut Support Person, a.k.a. Cape Crusader

New astronauts don't get a flight assignment right away. You might have to work for several years in support roles until a mission comes along with your name on it.

I volunteered for several jobs but ended up in Orbiter Systems. My section leader assigned me to work an issue with the shuttle's auxiliary power units. I wasn't exactly sure how to get started but eventually figured out that the job was basically whatever I wanted it to be. I worked on the issue with the APU controllers as well as problems with electrical, hydraulic, environment, and computer systems on the shuttle. I reviewed and approved changes to checklists. I answered questions for engineers and sat with them in the Mission Evaluation Room during missions.

Nine months later, Steve Nagel, the new chief of the Astronaut Office, asked me if I would like to be an astronaut support person, known as an ASP or Cape Crusader. At any given time, there were about five ASPs working at Kennedy Space Center (KSC). These astronauts were hands-on with the space shuttles and their crews, performing everything from tests in the hangar to securing the astronauts in the space shuttle for a launch.

I immediately started to say, "Yes!" but Steve stopped me. He told me to think about it for a day and discuss it with my husband. He knew Pat was on the road several days a week as an airline pilot. As an ASP, I'd be at KSC in Florida three or four days a week. Steve wanted to be sure that we were both okay with the two of us being on travel so often, which I deeply appreciated.

Of course, Pat was excited for me and was fine with my accepting the role. Our travel schedules would overlap, and we had no children to worry about. It was an easy yes.

ASPs performed tasks at KSC that required a trained astronaut. The actual flight crew was focused on their upcoming mission, so the ASPs took care of other details to ensure that the vehicle and the crew were ready to go. I ran flight control tests on the shuttle inside its hangar, setting pressures and testing the controls. We were integral to final preparations like the airlock closeout, the Crew Equipment Interface Test, and the Terminal Countdown Demonstration Test, which was a dress rehearsal of the launch countdown. As members of the KSC closeout crew, we would strap in the crews before launch and help them out of the space shuttle after it landed.

I also met and worked with various employees of the shuttle program. These visits with work groups gave me great confidence in the attitude and abilities of the KSC crews. They were extremely dedicated and serious about what they did. Whether in the tile shops, main engine installation, fire department, or a multitude of other

specialties, they were all groups of dedicated professionals performing important missions, and everyone felt a part of the bigger effort
of space exploration.

I worked on about ten missions during my time at KSC. The pre-
mier duty was assignment as the prime ASP for a mission, as a member of the seven-person closeout crew. The closeout crew wore suits
with different numbers on the back: number one was the team lead,
the prime ASP was number two, the quality control tech was number seven, and so forth. Closeout crew techs were highly trained, followed detailed checklists and procedures, and were all hand-picked
for the job. It was an honor to work beside them.

The closeout crew operated in the White Room at the 195-foot
level of the launch tower. We helped the astronauts don their harnesses and special equipment, and then we strapped them into their
seats aboard the space shuttle. This was an emotional time, when you
stop to consider that the closeout crew were the last people these
astronauts would see until they returned from space. The toughest
job was the physical effort required to close and secure the hatch,
which took plenty of strength and technical knowledge.

Although I had assisted crews on three prior missions, my first
time working as prime ASP was STS-50, the US Microgravity Laboratory 1 mission. I spent many hours "in the vertical" inside *Columbia*
on the launchpad, preparing the orbiter for its flight. The cockpit
looks like a whole new place when it's tilted back ninety degrees for
launch.

On the afternoon of June 25, 1992, we strapped in the crew, closed
the hatch, secured the White Room, and then drove off to a fallback
area three miles from the launchpad. At that point, there was nothing much we could do except hope that we had not made a mistake.

At T-31 seconds, when the onboard computers took control, there
was quiet anticipation, and (I am sure) much praying.

The first visual evidence that a launch was about to happen was
the water deluge, beginning at T-12 seconds. About 300,000 pounds
of water pours out of "rainbirds" onto the mobile launch platform

and down into the cutouts for the engines. NASA installed this system after the first shuttle launch, when the enormous acoustic energy produced by the space shuttle's engines and boosters bounced off the launch platform and damaged the launchpad and the shuttle's tail end. Some smart engineer figured afterward that a massive deluge of water would suppress the sound waves, thereby limiting any physical damage to the delicate space shuttle.

The shuttle's main engines fired at T-6.6 seconds. At T-0 the solid boosters ignited, and we observed liftoff, but we didn't hear any launch sounds until about nine seconds later. That's of course because it takes the sound five seconds to travel a mile. Distance and the speed of sound also delayed the ground rumbling and the air pulsations from shock waves.

I was amazed at the pure intense brightness of the exhaust. Shortly after liftoff and up until the time the shuttle rolled over onto its back, the visual brilliance was simply breathtaking. Cameras just can't capture the way it looks.

We held our breath until the shuttle reached MECO (main engine cutoff) about eight and a half minutes later. Then a great feeling of success washed over us. Our job was over until *Columbia* and its crew returned to KSC on landing day, scheduled for nearly two weeks later. We exchanged high fives and hugs.

My second mission as a member of the closeout crew was STS-47. This was one of the more unusual shuttle missions. It was the Japanese Spacelab flight, and it was the second flight of *Endeavour*—the newest shuttle in the fleet, built to replace *Challenger*.

The most memorable part of the mission was the crew. Mamoru Mohri was the first Japanese national to fly in space. Dr. Mae Jemison was the first African American woman astronaut. The first married astronaut couple, Mark Lee and Jan Davis, were also on this flight. NASA has a policy of not assigning married couples together, but Mark and Jan had secretly wed during their training. NASA found out too late to remove either of them from the crew before the launch, so they flew together.

With all the interest in the crew, STS-47 contributed greatly to the positive image of spaceflight. I attended the crew press conference at the launchpad three weeks before launch, and there were hundreds of reporters present. Flight opportunities were opening to more people of all nations and backgrounds. The public was also intrigued by the unique science experiments aboard—Japanese koi fish, chicken embryos, fruit flies, plant seeds, frogs, and hornets.

Since this was such a high-profile mission, Vice President Dan Quayle attended the launch with his wife, Marilyn. Mr. Quayle wanted to talk with Mae and Mamoru during the launch countdown. It was an unusual request, as normally only the launch director's representatives communicate with the crew during countdown. However, the launch team put Mr. Quayle on the comm loop during the T-9 minute hold. Mr. Quayle's comments went on for way too long, and the launch control team started to worry about launching on time.

I had the opportunity to meet the Quayles when they walked through a receiving line after *Endeavour* was safely in orbit. While this was a big morale boost for employees, the launch director subsequently decided that a politician would never again interrupt the process of a launch countdown.

My third and final time as prime ASP was STS-55, the German Spacelab flight. Everything went fine during strap-in, and we fell back to the fire department building as planned. *Columbia*'s three main engines ignited at T-6.6 seconds and then abruptly shut down at T-3 seconds. The launch control team called, "Cutoff!"

This was a serious situation. We did not immediately know why the engines had shut down. We had to prepare to return to the launchpad on a moment's notice to quickly extract the crew and evacuate the pad if necessary.

Our closeout crew leader quickly called us all together and reviewed the hazards. For example, there could be a hydrogen leak, which would be a potentially explosive situation. A similar engine shutdown on the pad on STS-41D in 1984 resulted in a hydrogen fire around the base of *Discovery*. A fully fueled space shuttle was potentially a six-million-pound bomb on the launchpad.

The NASA Test Director announced "all clear" and sent us back to the pad, where the ground was flooded with water. After STS-41D, it was standard procedure to activate the fire suppression system and douse the aft end of the shuttle with water if the main engines had shut down unexpectedly.

We took the elevator back to the 195 level, opened the hatch, and helped the crew exit. They were in a surprisingly good mood, considering they knew it would take two to three months to replace *Columbia*'s engines before the next launch attempt. I was sure the crew was disappointed, after months of training and coming so close to a launch, and now having to go through it all over again.

When we returned with everyone to the suit-up room at the crew quarters, I was shocked to see blood running down one astronaut's arm. I learned that he was a test subject for a research experiment during launch, with an IV inserted into an arm vein before he suited up. At some point during the day, that IV had pulled out. It would have been a mess to deal with in orbit!

Throughout my sixteen months as an ASP, I worked in a support role on seven other missions. We would watch launches with the crews' immediate families, their extended families, or VIPs in various locations. Our presence was vital for these people, as many attendees were understandably excited, nervous, or just plain overwhelmed by the emotions of launch day. We wore our blue flight suits and mixed with the crowds, answering questions and calming nerves. We weren't as close to the launchpad as the closeout crew, but still quite near enough to experience the physical thrill and power of a launch.

Working at KSC gave me an appreciation for the teamwork and the incredible feeling of success one has when a mission launches. I shared the sense that each and every person who works on a shuttle mission feels when it blasts off, that you are truly a part of the mission. The reward of being a part of the space program goes beyond the satisfaction of seeing the fruits of your labors launch into space. All participants know they're part of something bigger: the human need to explore the unknown and the ultimate need to expand human presence beyond the surface of the Earth and into outer space.

Time to Move On

The months from February 1992 to June 1993 provided me valuable experience at KSC, but sixteen months was a long time for a rookie pilot astronaut to be away from practicing for her own eventual mission assignment. My flight skills in the simulator had grown rusty.

That became obvious when I participated in an "integrated sim" with Mission Control in spring 1993. We simulated four ascents, each of which ended in a different kind of abort. The instructor team injected malfunctions into the simulation—sometimes two or more unexpected situations per minute. The commander and pilot had to prioritize and run multiple emergency procedures from the checklists while communicating with Mission Control, who used the sims to certify their flight controllers.[*]

I couldn't keep up with all the malfunction procedures. I wasn't throwing switches fast enough, and my calls to the ground were late. "If you don't use it, you lose it!" as the saying goes. I certainly wasn't proud of my performance that day.

After that experience, Hoot Gibson, the chief astronaut, said it was time for me to come back to Houston full-time. He moved me into a CAPCOM role. The CAPCOM (short for "capsule communicator," a term from the Mercury era) is the astronaut who sits in the Mission Control room and talks to the crew in orbit.

Here in Houston, I could work on operational issues—particularly emergency procedures—while fulfilling my training requirements in

---

[*] Pilots do less of this as technology matures and systems become automated. Computers and artificial intelligence on modern spacecraft can diagnose many problems and resolve them automatically, without the pilot's involvement.

the sim. My proficiency quickly returned. Although I missed my KSC job, I knew I would be back there someday—for my own launch.

A Punch in the Gut

Ever since completing ASCAN training, I had been waiting for my own mission assignment. The space shuttle routinely carried a crew of seven astronauts. Five of those seven were mission specialists or payload specialists, but there was only one pilot and one commander, and you could be a commander only if you had flown at least once— and usually twice—as a pilot. So, for me, the pipeline was long and narrow. I knew it might be several years before my first flight.

How many people were waiting for a ride? When I finished my ASCAN training, there were nearly forty active commanders and pilot astronauts. Six pilots were waiting for their second mission. Twenty pilots had not flown at all, including six of my ASCAN classmates and me. We were all vying for scarce, coveted seats, as NASA flew only six to eight shuttle missions every year.

In 1992, NASA announced STS-61, a daring mission to repair the Hubble Space Telescope. Mission specialist Story Musgrave was the first astronaut named to the crew. I was in the crew quarters at KSC later that year when the deputy of Flight Crew Operations gave me some great news. "We sent your name to headquarters for approval. You're on the Hubble repair crew!"

I was *so* excited! I couldn't possibly imagine a better first flight. He asked me to keep it quiet, so I didn't tell anyone except my husband. At KSC, my classmate Leroy Chiao told me that he was also informed he'd be on the mission. It was hard to contain our excitement for two months while we awaited the official announcement.

I was in my office in Houston on December 8, 1992, when astronaut Rick Linnehan popped by to say, "They announced the Hubble crew in the conference room." My heart stopped. The crew announcements were fun occasions, when the boss read out the names of the crew for the next mission or several missions. If you were on the crew, they would *always* tell you in advance of the formal announcement so you could be at the meeting. No one from management had spoken to me about it. I was massively confused.

"Who's on the crew?" I asked.

After he told me, I stammered, "I was told *I* was going to be on that mission."

I was absolutely crushed. I have never felt so devastated.

It took me a while to gather my courage and composure to call the deputy director and ask, "What happened?"

He sounded perplexed. "What do you mean?"

"You told me I was on the Hubble crew, but I wasn't on today's announcement."

Silence. Then he muttered something like, "Headquarters wanted a flown pilot." He said the mission had such high visibility that they didn't want any rookies on the flight, in any role.

I had to be careful how I handled myself in this conversation, because he was two levels above me in the chain of command. I asked, "Couldn't you have called me?" I imagine he felt horrible about what I was going through. He had simply forgotten he told me I would be on the flight.

I later talked to the mission commander, who was aware of what had happened with the crew change. It wasn't up to him to explain the politics to me, but he was kind enough to do so anyway. He told me that Mark Lee was also on the initial crew list and replaced by headquarters, and Leroy Chiao lost his spot because he was a rookie.

I found Mark Lee and talked with him about the situation. After we griped about it for a few minutes, he finally said, "I've got to move on." He was right. It was hard to let go, though.

I must be honest: I was very, *very* disappointed with everyone in the chain of command. I felt disheartened, like I was just a warm body that managers could swap out without any consideration about how it affected me as a person. To hear I would be on a high-profile mission, only to have it reversed without notice or explanation—it was almost too much to bear. It took everything I had to bite my tongue.

I tried my best to stay positive, continue to work hard at training, and focus on my job.

No matter where you work, your leaders and your colleagues notice how you deal with disappointment when you get passed over for an assignment or a promotion. Did you suck it up and keep doing your best, or did you complain to everyone who would listen? People remember your reaction to bad news, and it can affect future decisions about your career.

This is also an important lesson for leaders: If you make a promise to your staff, follow through on it, or keep them informed if you can't. Don't let your people be surprised by hearing bad news from someone else.

Hoot Gibson, who took over as chief of the Astronaut Office, found out what had happened. He told me, "Eileen, I'm going to put you on a *really* good mission." I'm forever grateful to him for lifting my spirits.

A few months later, Hoot called me to his office. I hoped he would tell me of a flight assignment. Instead, he said, "We're going to announce several missions this afternoon, but you are *not* on any of them. I'm just telling you this because I want you to know we haven't forgotten about you. I want to put you on a mission that's better than any of these! It will be a really, really good mission. You just need to be patient."

I said, "Thank you very much. That's all I needed to know." At this point, I didn't care which mission I was on. I was just happy to be treated like a person, not a commodity.

Waiting to find out about your first flight is nerve-racking. You feel happy for your classmates and friends assigned to crews, but you can't help wondering, *When is it going to be my turn? I worked all of my life to get here. Why won't they let me fly?*

Hoot was right. They eventually did give me a really good mission as my first flight assignment. But I had to wait an agonizing nine months after the STS-61 fiasco before NASA finally announced it.

Chapter 10

FIRST WOMAN SPACE SHUTTLE PILOT:
STS-63

In September 1993, Hoot Gibson told me I would fly STS-63, a SPACEHAB mission scheduled to launch in May 1994. I felt elated to finally celebrate a crew assignment!

Jim Wetherbee, who eventually went on to command a record five shuttle missions, was our commander. Bernard Harris was MS 1 (the lead mission specialist) and payload commander, in charge of coordinating all of the SPACEHAB experiments. British-born Mike Foale was MS 2. My Hairballs classmate Janice Voss was MS 3. Janice was so brilliant that our class nicknamed her Data after the android on *Star Trek*.

Our sixth crew member was cosmonaut Vladimir Titov, who would become the second Russian to fly on a space shuttle. Volodya, as we called him, was a happy and heartwarming man with nerves of steel. He had survived a September 1983 incident when the launch escape system pulled his Soyuz capsule safely away from a booster that exploded on the launchpad.

Jim's first instructions to me were to memorize four critical procedures, including dealing with a main engine helium leak or shutting down one of the auxiliary power units in the event of an emergency. Even though I would have the checklist in front of me during a mission, memorizing those procedures could save precious seconds and possibly our lives. You can multitask if you have things committed to memory. It improved our response time in simulator runs when the instructors threw malfunctions at us.

The press release introducing our crew to the public in September 1993 only announced the SPACEHAB laboratory that we would carry in the shuttle's payload bay as our primary mission. However, we heard that the United States and Russia were negotiating a plan behind the scenes for our space shuttle to approach—and possibly even dock with—the Russian space station Mir. Our compressed schedule gave us less than a year to train for the mission.

We quickly ran into a snag. SPACEHAB was woefully short of commercial and academic experiments to fill its lockers, and NASA didn't want to fly it half-empty. So, only one week after the press announcement, NASA pushed our mission back to late January and then early February 1995. This provided time for the United States and Russia to finalize the agreement to rendezvous with Mir. NASA announced in December 1993 that our mission would now be the first space shuttle flight to approach Mir.

Hoot was right. This would be an interesting, challenging, and exciting mission.

Delayed Mission, Delayed Family

The timing was lousy, though. A delay of nearly a year caused us to jump behind five other space shuttle missions on the launch manifest.

It also meant another year before I'd be able to start a family. Mike Foale came into my office one day and announced that his wife was pregnant. I told him how happy I was for them, although it demonstrated the difference between raising a family as a male astronaut or as a female. Despite our mission's delay, Mike and Rhonda could still have their baby. I would have to put off any thoughts of having children for now.

During pregnancy, women astronauts cannot fly in space, and certain training activities are off-limits. Most astronauts are in their early to midthirties when they join the program and are often in their late thirties by the time they begin flying. That's approaching the age where we worry about possible complications or health issues for our babies and ourselves.

This was a common topic of discussion for some of us. There is nothing you can do if your mission keeps delaying, and delays were common. Between missions, which could normally be several years, a woman could take herself out of the rotation temporarily. Or do you keep putting a pregnancy on hold, in hopes of a mission assignment sooner rather than later? There are no easy answers.

Pat and I wanted to start a family. We also knew how important my flying this mission was for me, NASA, and the women around the world who were following my career.

Despite my deep disappointment, I gradually decided to try to make a good thing out of the bad news. As is my usual practice, I worked through my frustration by mentally turning the tables: I tried to see this unhappy development as a gift of extra time to make myself the smartest, most knowledgeable pilot in the Astronaut Office. I would memorize the emergency procedures. I would sign up for more simulator time. I would fly more T-38 missions. I would study, study, study. If a malfunction happened, I would be the hero who would save the shuttle! On my long exercise runs, I imagined myself and my crew in all kinds of dire situations: fire in orbit, air leaks, cooling system leaks, engine failures. You name it, I would be ready for it. I would apply my chair flying skills to all kinds of dreamt-up situations, and of course my crew and I would come out as the heroes.

Working hard and visualizing success helped me deal with the heartbreak of the long delay. In the end, what had initially seemed to be a career disaster turned out to be the best thing that could have happened. I was perfectly prepared.

Space shuttle pilots spend long hours practicing both in the simulators in Houston as well as in the pilot's seat of our Shuttle Training Aircraft (STA). The STA is a Gulfstream II specially modified to fly like the space shuttle during its final approach and landing. Obviously, we can't practice supersonic atmospheric reentry in a real spacecraft, but the shuttle's computers handle that part of the flight. Humans take control in the final several minutes of flight, after the shuttle has become subsonic and has dropped below fifty thousand feet. We practiced this part of the flight in the STA.

The shuttle is falling like a brick at that point, shedding one thousand feet of altitude every few seconds. The commander must fly around an imaginary *heading alignment cone* and then line up with the runway, descending seven times steeper than a commercial airliner. Just before landing, you arm the landing gear at three thousand feet, pull back on the stick to "pre-flare" at two thousand feet to slow the descent rate, deploy the gear three hundred feet above the ground, and cross the runway threshold. The shuttle should touch down at either 195 or 205 indicated knots (214 or 226 miles per hour), depending on weight. It all happens so fast that the commander and pilot must build almost instinctual habit patterns as a baseline for dealing with unexpected adjustments like wind shear and turbulence. This was why the STA practice runs were so important.

To be qualified as the space shuttle pilot, you must fly at least five hundred approaches and landings in the STA. Commanders must fly one thousand of them. There is no substitute for this kind of real-life training.

I took every opportunity available for stick time on the STA. Flight test experience came in handy here. The instruments aboard the STA recorded all your approach and landing measurement data. You immediately received a printout of the critical information to review while the "safety pilot" took the plane back up to altitude for the next go-around. How fast were you going when you crossed the threshold? How far down the runway did you touch down? How close were you to the centerline? How fast were you flying when you touched down, and at what descent rate? You could see a plot of your trajectory and how your inputs on the controls affected your glide path. When a crosswind gust hit, did you overcorrect? Something invariably happened on every flight that presented you with an improvement opportunity—to try to come as close as possible to the elusive gold standard of an ideal landing.

The information and practice sharpened my skills. I shared my observations and the lessons I learned with my fellow pilots. We could all improve our proficiency by learning from one another, and that meant being open and honest about our performance as individuals. If I made an unusual error that caused something unpredictable to happen, I shared that information with my colleagues so they wouldn't make the same mistake. Humility and vulnerability were important. Too many lives were on the line if you were too proud to disclose what you learned from your errors.

Another result of our mission delay was that I'd be the next-to-last person in my astronaut class to fly. On occasion, that agitated me. When it did, Jim Wetherbee reminded me that Neil Armstrong was the last person from his astronaut class to fly. The first man on the Moon! Rob Navias from NASA Public Affairs reassured me, "Your mission will go on time in January 1995." He seemed confident, and his optimism was contagious.

Both men helped me to realize that there was no use in me stewing about things over which I had no control. I couldn't influence the launch schedule. I *could* control my knowledge and abilities.

Yes, there was a possibility that I might be too old to have children when I finished this mission at age thirty-eight, but I had signed up to be an astronaut, to fly in space. That was my lifelong dream, and I was going to fulfill it.

I would not just do my job—I'd be the most capable and knowledgeable pilot in the program!

Shuttle-Mir

Our crew flew to Moscow in the summer of 1994 for ten days of training with the Russians. Since we wouldn't actually be docking with Mir, we didn't need to train on the operation of their space station's systems. We were there primarily to get to know the people and learn about the culture.

As an officer of the US Air Force, it was hard to fathom being in the capital city of the country that was our primary adversary just a few short years earlier. Russia was still recovering from the economic upheaval after the breakup of the Soviet Union. Our guides privately apologized that Moscow was not as clean and majestic as it once was.

We stayed at the Penta Hotel and drove to Star City on the northeast outskirts of town for our training. Star City's facilities showed some signs of neglect. The pool in which the cosmonauts practiced spacewalks was so dirty that it was unusable. Some buildings had no lights. Lenin's photo still hung on the wall of our training room.

I enjoyed sightseeing in Moscow during our free time. We visited Lenin's tomb and Red Square. We toured several aircraft and spacecraft museums. We went into the cockpit of their *Buran* spacecraft, which was the Soviet version of the space shuttle. (It only flew once—for one orbit and without a crew—before being mothballed as too expensive to operate.) We met Volodya Titov's wife and daughter, who warmly welcomed us.

We were excited to be working with the Russians, and we also knew that progress would not be easy. Everything, and I mean *everything*, required negotiation. It would take several joint flights for both countries to feel comfortable about the space shuttle attempting to dock with Mir. The first step would be to fly a cosmonaut, Sergei Krikalev, on the space shuttle. He would communicate with Mir by radio in orbit. Since our flight would approach Mir in orbit but not dock with it, our mission acquired the nickname "Near-Mir." The first actual docking would occur on mission STS-71, four months after ours.[*]

Once the Russians agreed to our mission's rendezvous with Mir, we began discussing how close we could come to their station. In the spirit of building up capabilities, could we fly the space shuttle within one thousand feet of Mir to test out the flight qualities, navigation systems, communications, and crew and mission control procedures?

---

[*] Space shuttle mission numbers came from the order in which the missions were initially assigned, not the order in which they actually flew. When delays caused missions to be moved around in the launch schedule, the flights might not be in numerical order.

Yes, the Russians agreed that it seemed logical. We began developing the procedures and training for a one-thousand-foot approach.

As we grew more confident, it became obvious that we could test procedures and eliminate additional technical uncertainties by coming even closer to Mir. That would decrease the risk for the STS-71 docking. I watched Jim Wetherbee and his Russian flight director counterpart Viktor Blagov negotiate through an interpreter, while standing in the historic control room at the Korolev mission control center. (Everybody called it the "TsUP.") They eventually agreed that a rendezvous at four hundred feet was acceptable.

After training for that approach, we thought, "Why not fly even closer, say, one hundred ten feet (thirty-three meters)?" That was negotiated and eventually agreed. Then we wondered, "Why not thirty-three feet or ten meters?" Ten meters was the distance at which a space shuttle would begin its final docking maneuvers. We could test everything except the docking itself.

Discussions became much more intense to persuade the Russians to agree to such a close approach. Their primary concern was our flying a one-hundred-ton spacecraft so close to the pride and joy of the Russian space program. The Russians weren't convinced that we could precisely control the space shuttle if anything went wrong. The Russians reminded us about Deke Slayton's bumping his Apollo spacecraft (although this is disputed) into the Soyuz back in 1975. Nevertheless, everyone agreed on the final distance: ten meters.

I learned a little about how Russians negotiate. It seemed to me that they basically wanted the same thing we did. They started discussions by making you think they *didn't* want it. That way, they got you to make concessions, and you felt like you made more progress. I also learned that it was difficult to believe anything you heard. You could ask the same question of three different people and get three different answers.

Even though the first two women in space were Soviet citizens, only four female cosmonauts have flown in the entire history of their program. I don't believe the Russians treated me any differently because I was a woman or that they were uneasy about having a woman in the pilot's seat on the flight. My impression was that the Russians deferred to the commander for questions and decisions. It was simply a matter of talking to the person in charge.

If you made a sincere effort to try to understand Russian culture and worked with them in that context, and you showed them the proper respect for their technology and accomplishments, you'd do fine. You got into trouble if you tried to impose your opinions or demand that they do something your way. Even though Jim didn't speak Russian, he was a superb diplomat. Once they got to know him, they trusted him implicitly.

Launch Preparations

As our flight date neared, we traveled to Kennedy Space Center for the Crew Equipment Interface Test—putting your hands on the actual hardware you will be working with on the mission, not just a training model or a mock-up. We checked out our experiments before they were stowed on board *Discovery*. In Houston, we packed our food, clothes, in-flight maintenance kit, and anything else we might need for a medical emergency in space—IVs, medication, and dental equipment, for example.

Every astronaut must test about twenty different types of medication for possible medical issues that might occur during a mission—everything from aspirin to sleeping pills in varying strengths. Beginning about a year before your mission, the flight surgeons ask you to try each of the medications (one at a time) to check for any adverse reactions. Better to discover an allergy on Earth than to find out in space! I tried them all—everything except a tranquilizer called Halcion, used for severe insomnia. I knew I would never need that one. As it turned out, I never took any sleeping medication on any of my missions. If we had to make an emergency de-orbit and landing, I didn't want anything in my system that would cloud my mind or deaden my reflexes.

I felt incredibly enthusiastic about our mission. We had such a diversity of important activities and milestones! Bernard Harris would become the first African American to walk in space, and Mike Foale would join him as the first British-born person on an EVA. We'd have the second Russian to fly on a space shuttle and the first woman in the pilot's seat. We'd be the first Americans to see Mir in orbit. We would launch and later retrieve an astronomy satellite called SPARTAN, and we'd have dozens of scientific experiments in our SPACEHAB laboratory.

But what made the headlines in the Houston newspaper after our preflight press conference? "ASTRONAUTS TO FLY COKE IN SPACE."

We flew to Kennedy for the Terminal Countdown Demonstration Test on January 17, 1995. After my many times at the Cape in support roles and working on closeout crews, now it was *my* turn to be strapped into my seat on *Discovery* by a fellow astronaut. We went through a simulated countdown and practiced our emergency escape procedures. Then it was back to Houston for our final simulator runs and preparations for the mission.

Dr. Sally Ride called me the day our crew went into our week-long quarantine. As the first American woman to fly in space, she'd had more than her share of press attention. She had endured hundreds of questions about what kind of makeup she'd be wearing, while almost no one asked about the experiments she would be performing in orbit. Sally told me, "Don't be distracted by all the silly media questions. Just focus on your job."

That same day, the Rev. Jesse Jackson phoned Bernard Harris. He said, "You are taking the Black man from the slave ship to the spaceship!"

While observations like these certainly put our roles into historical perspective, Bernard and I just wanted to *fly*. We only wanted to perform well so that our mission would be a success. We knew that the best way to honor our places in history was to be our absolute best at our jobs. Reveling in the spotlight would be a disservice to all the people who had come before us and who enabled us to be where we were now.

Astronaut Story Musgrave gave me an interesting suggestion about how to prepare myself for life in space while I was in quarantine, beginning seven days before launch. In orbit, your sinuses feel full and stuffy because gravity isn't pulling blood down from your head. He said I should try "heads-down" sleeping to help me acclimate to that feeling. To raise my feet above my head, he told me to put several bricks under the foot of my bed, which I found quite uncomfortable. I kept sliding out, so I quickly gave up. It was more important to get a good night's sleep and feel rested for launch.

We flew our T-38s to Kennedy Space Center on January 29. Some important guests were waiting to view our launch—seven of the surviving members of the Mercury 13, also known as the FLATs (Fellow Lady Astronaut Trainees). These highly skilled women pilots had volunteered in the early 1960s for a private program to see if they could pass the same physical tests as the Mercury astronauts. Of the twenty-five women who tested, thirteen of them equaled or bettered the performance of the men. They could have qualified for the astronaut program, except for one technicality. NASA at the time required all astronauts to be test pilots, and women weren't permitted to fly in the military, much less attend test pilot school.

The Mercury 13 and I shared the bond of being pilots who yearned to fly in space. I wanted them there for my launch. If they couldn't go into space themselves, I wanted them to know that I was flying on their behalf. They made it possible for me to achieve my dream.

Two days before our scheduled launch date, and while we were still in quarantine, NASA held the traditional beach house party for the crew, our immediate families, a few members of the launch team, and a couple of special guests. The astronaut beach house was about four miles down the beach from Pad 39B, where our ship *Discovery* awaited countdown. It was the perfect place for everyone to get together out of the eyes of the media and enjoy some fellowship with one another.

My parents came, as did Pat and his parents. Volodya Titov brought a special guest: Valentina Tereshkova, the first woman to fly in space. I'll never forget the thrill of meeting her. She was only twenty-six years old when she flew in 1963—and here I was, thirty-eight years old for my first flight. She spoke only Russian. Volodya acted as our interpreter. I wish I could say that we had a profound conversation about the importance of being women space pioneers. Instead, it was more along the lines of "How nice to meet you! Welcome to America! What do you think of Florida?"

The six people on the crew and all of our invited guests filled the house. The cooks prepared a wonderful barbecue meal. With all the Russians present, we were toasting every person, place, and thing all night long! My parents got a big kick out of that and talked about it for years afterward.

All too soon, it was time for everyone except the astronauts and their spouses to leave. The extended families took a bus back to their hotel. We spent a last hour saying goodbye to our spouses before they too had to head back to the hotel. Then we retired to crew quarters at the Operations and Checkout Building.

Launch Countdown

Our launch would be just after midnight Eastern Time on February 2. Because of our late-night liftoff, we had been shifting our sleep schedule so that we would go to bed at about eleven in the morning. The daylong countdown was already underway when I went to bed. Surprisingly, I slept soundly.

Just before we were supposed to wake up during the evening of February 1, Bernard Harris knocked on my door: "Go back to sleep. We're scrubbed for today." I fell right back to sleep.

One of *Discovery*'s three inertial measurement units had failed and needed to be replaced. Those units contained gyroscopes and accelerometers that told our computers about our orientation and acceleration. Technicians could swap out the units while the shuttle was on the launchpad, but it was a mandatory twenty-four-hour turnaround.

When you're planning a rendezvous with a space station in orbit, launch timing is tightly constrained by the laws of physics and the amount of fuel we could carry. The Earth is rotating underneath the orbit of the space station. The space shuttle has only enough fuel to catch up with an orbiting space station at the moment the station's orbit passes directly over the launch site and when the space station is a certain distance ahead of the launching shuttle. We had to catch up with Mir at a precise time and location above the Earth several days later. What it meant to us astronauts planning to rendezvous with Mir was that we only had a five-minute "window" each day in which to launch.

Launch scrubs were commonplace in the space shuttle program. They were mostly due to weather at the Cape or the emergency landing sites across the Atlantic. Equipment or software issues caused delays, too. We learned not to let it bother us. At least we hadn't suited up and strapped in at the launchpad only to have to turn around and come back again. I spent the day reviewing procedures, reading a book, talking with my crewmates and the support astronauts, and following the emails about the status of *Discovery*'s repairs.

Things looked good for our next launch attempt, now set for 12:22 a.m. (EST) on February 3. It would still be February 2 back home in Houston, so I always have to explain that we launched on February 2 or 3 depending on where you lived!

At our crew breakfast in the evening of February 2, I only had a piece of toast and some water. After hearing so many stories about rookie astronauts getting space adaptation syndrome (space sickness), I overcompensated and didn't eat enough. I probably would have been better off eating eggs and bacon like the rest of the crew.

The suit technicians helped us into our orange "pumpkin suits." Then we walked down the hallway, rode down the stuffed elevator, and strode out to the waiting Astrovan. Our training team came from Houston to cheer us on, and they were hooting and hollering as we left the building. What a great group they were! They flew to KSC on their own time and at their own expense to see us off.

I didn't realize that Jim Wetherbee was closely watching each member of his crew as we took that long ride out to the launchpad.

He later said that as the commander, he wanted to see how his crew was dealing with the stress before launch. He said that as we turned the corner and confronted the beautiful sight of *Discovery* bathed in the brilliant xenon lights, he could instantly tell that we were "psychologically, intellectually, and emotionally ready to operate at an elite level."

The six of us and a member of the closeout crew crammed into the "slowest elevator on Earth" and rode to the 195-foot level of the launch tower. Jim and Volodya were the first to walk across the access arm to the White Room, don their parachute harnesses, and strap into their seats on board *Discovery*. The rest of us stood on the open grating at the service tower and waited our turn.

A fully fueled space shuttle was definitely a living, breathing machine. Pumps constantly added liquid hydrogen and liquid oxygen to replace what was being boiled off as the massive vehicle waited for launch. Even in February, the Florida air was humid. The water vapor condensed onto the cold fuel tank and then dripped off like rain. Oxygen vapor loudly puffed and hissed from vents on the side of the tank, creating an eerie mist in the bright spotlights. The wind whistled through the open trusses and grating of the launch structure. Imagine walking through a cemetery on Halloween night as the fog is rolling through!

I was the third of our crew to cross the access arm. Once I was in the White Room, it started to feel a little less surreal, because the lighting was normal, and I was not outside in the elements.

Since the shuttle was nose-up, I had to duck into the hatch, turn right, and crawl across the platform that covered the ladder between the mid-deck and cockpit. Then it took a good deal of upper body strength to twist, climb onto the back of the MS 2 seat, and pull myself up into the pilot seat while wearing my eighty-pound pressure suit.

The cabin lights illuminated the shuttle's cockpit. It was too dark outside to see anything through our windows except the glare of the spotlights.

Having worked so many missions as a support person helped me feel comfortable and confident with the process. Astronauts used to joke that people who had been Cape Crusaders were the worst crew members to strap in, because they'd done the task before and would constantly be telling you how to do your job. I tried to keep that in mind and respectfully let my colleague do what he had been trained to do.

Welcome to Space

I've already told you about the launch and wild ride to orbit. Eight seconds after our main engines shut down, the computer fired the explosive bolts to separate *Discovery* from its large external fuel tank. A few of the shuttle's thrusters fired to move us away from the tank.

At that instant, the master alarm sounded. The *doo-doo, doo-doo, doo-doo* and flashing red light caught my attention! I punched the spring-loaded square button to silence the alarm and checked the computer monitor. One message said JET LEAK, and there were two JET OFF indications.

Some of our maneuvering thrusters were failing.

One of the failed thrusters was R1 Upper (R1U for short). It was an upward-facing thruster on the pod on the right-hand side of our tail. As we opened the payload bay doors, Mike Foale looked out the rear cockpit windows. "We really have a leak! It looks like a geyser!"

The rest of us groaned, "Mike, don't say that!" That jet was leaking in the exact direction where Mir was going to be when we made our close approach. It wasn't just a small leak. You could follow the conical plume of oxidizer shooting off five miles into space above the shuttle.

There are forty-four primary and vernier thrusters on the space shuttle. Some, like another leaking one facing downward on the port side of the tail, we could "deselect" and forget about. We would need R1U for the Mir rendezvous, though.

The Russians wouldn't permit us to come anywhere close to Mir if we couldn't fix the problem. We would risk contaminating Mir's solar panels and the optical sensors on the Soyuz spacecraft docked to the station. Our mission rules required that all of the thrusters at the aft end of the shuttle be working if we were going to approach closer than one thousand feet to Mir.

I clumsily floated out of my seat, banging into the control panels. I took off and stowed my pressure suit as quickly as I could and clambered back into the pilot's seat.

The crew was agonizing over possibly losing the primary objective of our mission. I wasn't feeling great, either—and it wasn't because of the thruster. I had never, ever been nauseous in an airplane, no matter which dizzying acrobatic maneuvers I performed. Previous astronauts had reported that there was no correlation between being airsick in a plane and being sick in space. Now, I felt queasy. I kept worrying, *Am I going to get sick?* Thinking about it certainly didn't help.

I asked Bernard Harris, who is a medical doctor, if he would give me an anti-nausea shot of Phenergan. He looked at me and said, "Nah, you're fine. Try taking a pill."

That was not what I wanted to hear.

My other physical problem was that I felt a constant need to urinate. It's as if a person has a microswitch at the top of their bladder. When you're on Earth and that microswitch gets wet, it's your signal to go to the bathroom. In space, everything is floating around, so that "sensor" at the top of your bladder is always wet. I felt like I had to go *all the time*. It seemed like I was in our bathroom every fifteen minutes the first few days of the mission.

While that annoyed me and interrupted my schedule, it was also potentially dangerous to my health. Astronauts can quickly become dehydrated in space, and I felt so queasy I couldn't think about drinking or eating.

We completed our first day's activities, which mostly involved setting up equipment. At the end of our day, I tied my floating sleeping bag to the ceiling of the shuttle's mid-deck. I couldn't sleep. My mind kept replaying the thrilling launch and ascent.

The morning of our second day in orbit, my task was to assemble the bicycle ergometer in the mid-deck. We had trained it as a two-person job, but the other crew members were busy, so I did it alone. Things floated everywhere around me. I lost my sense of orientation. I'd float and rotate while I concentrated on assembling a part, not noticing that I'd moved. I'd look up, but the ceiling was no longer above me. My brain was getting crosswise messages about my location. Since the components in my inner ear were floating in zero gravity, they couldn't provide me with any sense of up or down relative to what my eyes were telling me. I felt completely disoriented and then terribly sick. I had the dry heaves, as there was nothing in my stomach. I sipped a little water, and it came right back out again. I was unable to complete the setup without a remedy.

This time, Bernard gave me the Phenergan shot. Within seven minutes, I felt fine.

It still took a few days to regain my appetite. I ate a few squares of Chex mix that day, and a few more on our third day in orbit. By the fourth day, I finally felt well enough to eat that night's dinner—chicken and rice with vegetables.

We continued to work on the R1U thruster during our second day in orbit. Houston and Moscow negotiated a backup plan with a new safe distance for us to approach Mir if we couldn't correct the problem. We ran a test of our thruster system, and yet another jet began leaking—this one in *Discovery*'s nose. Its temperature dropped below its operating limits. We had to shut it down and then turn the shuttle so that our nose was in the sunlight. We cycled the valve a few times, and finally the thruster came back online.

Rendezvous with Mir

I was glad to be feeling better by the third day of the mission, because we were rapidly catching up with Mir. We were about to get busy. We shut off the manifold that supplied oxidizer to the R1U thruster and "baked" it by pointing *Discovery* to warm the jet nozzle in the sunlight.

When we woke up on flight day four, the R1U thruster was still leaking a bit, although not nearly as badly as it had been before.

Houston told us the wonderful news that we could approach within four hundred feet of Mir. Once we reached that point, flight controllers would decide whether we'd continue our approach or if we'd fly around Mir, staying four hundred feet away.

The laws of orbital mechanics prevent you from just pointing your ship at your destination and firing the engine like they do in movies. You'd completely miss the target! Rendezvous in space is a slow, deliberate, careful dance with no sudden moves. There are several different methods in which a vehicle like the space shuttle can rendezvous with a big target like Mir in orbit. All of them start with the space shuttle behind Mir and in a lower orbit. The lower your orbit, the faster you go. Today we would fly a *V-bar* (velocity vector) approach. That meant we'd fly under Mir and get ahead of it, then slow ourselves gradually so that we would climb to a slightly higher orbit, intersecting and matching Mir's orbit when we were about 1,100 feet in front of Mir.

We would use a different set of thrusters to slow ourselves down as we approached Mir. In a normal rendezvous, we would use the jets like R1U that directly faced in our line of motion. Since we didn't trust that thruster, we would execute a low-Z approach for the final phase. Firing thrusters mounted at an angle from our direction of motion would slow us down, but this less efficient pointing consumed nine times as much fuel as our preferred method.

In addition to working with our unusual thruster configuration, Jim and I had our hands full flying the approach. Jim needed to concentrate on flying precisely and evaluating the overall system. (After all, this was a test flight.) I called out the procedures, operated the radar, and anticipated and watched for malfunctions, prepared to take immediate corrective action if anything went wrong. Mike Foale operated the laptop computer running the rendezvous and proximity operations (R-POP) software program.

Three days after we closed the manifold for the R1U thruster, its contents had finally emptied into space, and there was nothing left to leak. We awaited word from Mission Control about whether we would be permitted to move in closer than four hundred feet.

Though we couldn't hear the conversations going on between Houston and Moscow, we had a direct radio link with Mir, and the Mir crew gave us the exciting news that we could approach within ten meters! Houston came on a few minutes later with the official confirmation.

Jim flew a precise series of maneuvers that put us in front of Mir, and then he moved us in to four hundred feet. We held there and waited for the final "Go for Approach" from Mission Control.

Mir was growing larger in our overhead windows. In the dark, we saw only its pulsating position lights. After sunrise, it looked like a bright dazzling cluster of metal cylinders, shining and glistening in the sun's reflection. Some of us compared its appearance to a dragonfly.

I ran the shuttle's systems that measured our range from Mir. The star trackers in *Discovery*'s nose gave us good positional information when Mir was far away and looked like a star. Our radar had no problem locking onto Mir from a distance. As we drew closer, though, Mir became much larger relative to the radar dish. The radar became confused and began feeding "ratty data" to the navigation computer. It would lock onto one of the modules, then wander to one of the solar arrays, and then it would find a docking port, giving us constantly changing figures for our range. Fortunately, we had a handheld laser rangefinder that we operated through the overhead windows at the aft of the flight deck. We could aim the laser more precisely than the radar, and it gave us the most accurate range information during the close approach.

Jim flew *Discovery* smoothly, and we stopped at the precise spot relative to Mir that we had planned. We stayed there for fifteen minutes while Jim and Aleksandr Viktorenko, the commander of Mir, exchanged words of fellowship to mark the occasion.

"As we are bringing our spaceships closer together, we are bringing our nations together," Jim said. "The next time we approach, we will shake your hand, and together we will lead our world into the next millennium."

I saw cosmonauts Elena Kondakova and Valeri Polyakov peering at us through Mir's portholes. (I didn't know then that Elena would be my crewmate on my next shuttle mission!) Valeri had already been aboard Mir for more than one year. When he came back to Earth on March 22, he had set a record for the longest spaceflight in human history—437 days—a record that still stands.

Jim slowly backed us off to four hundred feet and then executed a slow fly-around of the station. We fired our engines for our final separation from Mir, and, within a couple of hours, Mir was just a small flashing star off in the distance with the Earth down below us.

We received a call from President Clinton a few hours later. He offered his congratulations for accomplishing this first step in what was sure to be a long international partnership. He said, "People all over our country and all over the world will be seeing you today, and saying, 'This is something worth doing.' You've made us all very proud." We thanked him for his leadership and vision that were enabling the next great phase in manned space exploration: the construction of an International Space Station.

Living in Space
The day after the rendezvous, I finally had time to relax and think about something other than our flight plan. I was in the galley getting a drink, when suddenly it hit me. *Here I am, in space! Really, wow! I'm here.*

I thought about my grandmother, Marie Reidy Collins, who passed away in 1985 at eighty-nine years old. She never learned how to drive. I remembered taking her to the grocery store every Thursday afternoon while I was a community college student. She had a terrible time getting in and out of my car. It occurred to me, *If my grandmother were here in space, she could do everything! No gravity to battle. No need to worry about falling down.* That was always her biggest fear. Gravity is such a huge factor in our everyday lives on the surface of Earth, yet it's something we never think about unless we're afraid of falling or dropping something.

If old people could live up here, they would be able to do it all!

I learned how vitally important it was to keep clutter to a minimum. An astronaut has to be organized and know where all her stuff is. Trying to find a misplaced item while in orbit wastes a tremendous amount of precious time.

There are three ways an astronaut can lose things in even a small volume like the space shuttle. First, things will float away from you. If you let go of something, it had better be on a lanyard or attached to something with Velcro. Second, you can easily forget where you put things. On Earth, when you put something down, it will usually be on a horizonal surface like the floor or a tabletop, in the lower half of the room. In space, the entire volume of the compartment and all its surfaces are available. The walls and ceiling become places where you can stick things, and you can even attach small items to your clothing. There's no up or down, your orientation changes constantly, and your brain has trouble keeping up with the weird rules of the new environment while you're trying to execute the tasks on your checklist. Casually sticking a pen to a wall might mean you won't find it again for the rest of the mission.

The third way to lose items is when your crew members take them. Sometimes it was unintentional; sometimes it was a joke. Jerry Linenger pilfered my spoon for three days on my second mission. It was funny at first, but since I was issued only one spoon for the entire mission, the humor quickly wore off!

Managing trash was a big deal. You can't have garbage floating around inside the ship, so you'd try to minimize the amount of trash you generate, for example, by not completely cutting the top off a packet of food. Compacting trash and mashing it down into the bottom of a waste bag was an acquired skill, because you couldn't use your body weight to stomp things flat. You'd have to anchor your back against a wall, place the bag between your feet and the ceiling, and try to compress it as much as possible to get all the air out. Then you used gray tape to wrap and seal the bag.

Sealing any holes was especially important if the garbage bag held used food containers. The prime directive of food consumption on the space shuttle was "If you open a food package, someone has to eat all of it. *No leftovers.*" If you couldn't eat all of your food once you opened it, you offered it to your crewmates.

Imagine driving around in a minivan with a trash bag containing a week-old package of half-eaten shrimp cocktail, and you get the idea. The smell of some items, even when freshly opened, became repulsive in space. Some commanders forbade fish or bananas on their missions.

We stored the sealed garbage bags under the floor panels in the shuttle's mid-deck. In the old days of the program, some unfortunate technicians at Kennedy had to sift through those bags after the mission to ensure that the crew hadn't accidentally thrown out any tools or other equipment. Astronaut Marsha Ivins ended that practice, much to the relief of the shuttle processing teams.

Experiments in Space

Our SPACEHAB module and *Discovery's* mid-deck lockers held about two dozen scientific experiments. They ranged from an astronomy satellite in our payload bay to equipment to grow protein crystals in zero-g. Our four mission specialists were in charge of operating most of them.

I volunteered to run one unusual experiment that sounded like it would be fun: the Fluids Generic Bioprocessing Apparatus (FGBA), developed by the University of Colorado and funded by the Coca-Cola Company and several other firms. You probably know it better as the "Coke Experiment." NASA didn't allow us to call it that.

Coke and Pepsi had both flown specially modified soft drink cans on the shuttle in 1985. This time, we had a dispenser system similar to those in restaurants on Earth, with regular Coke on the left side and Diet Coke on the right. The dispenser mixed the syrup with carbonated water and fed it through a hose into a clear cylindrical container. Since there was no gravity, surface tension made the liquid stick to the inner sides of the container. Then we would drink the Coke and evaluate several qualitative factors on a questionnaire. On a scale of one to ten, how sour was it? How sweet was it? Did it taste the same as on Earth? We had to film the whole process from several angles and downlink the video live to Houston.

Bernard Harris and I set up the experiment on the fifth day of the mission, positioning the cameras and readying the paperwork. Jim Wetherbee floated into the SPACEHAB and asked, "What are you doing?"

"I'm setting up FGBA. We're ready."

He looked at the cameras and said, "Turn the cameras off. No live downlink."

I asked why.

"George Abbey said, 'No live downlink.'" Jim floated off to attend to some other task.

I turned the video feed off.

Mission Control immediately called up, "We need live downlink. Turn the cameras on."

What was I supposed to do? I couldn't disobey an order from my commander, who was in turn obeying an order from the deputy director of Johnson Space Center. Meanwhile, Mission Control was calling for me to send them video. And now, the payload control center—the people who had paid for this experiment—were also demanding video.

We ran the experiment and recorded video for the scientists to watch after the flight. We ran it several more times during the mission, again not sending the video to the ground.

What were my observations? Coke in space tastes like Coke on Earth. The truth is that I didn't want to drink it. I just never desired anything sweet when I was in space. I wasn't burning a lot of calories, so I didn't need the energy rush from sugar. I craved salt instead, because the human body has trouble maintaining its fluid balance in space.

I was surprised at the extreme media interest in our FGBA experiment but eventually realized that reporters like to cover anything their readers can relate to—food, drink, Coke. I just wish that the press covered more of our science work and its importance to people on Earth.

NASA endures endless complaints about "frivolous" activities like FGBA or video of astronauts playing with their food in zero gravity. People don't realize that these clips are the only ones that end up on TV. Astronauts in space spend the vast majority of their time performing far more important work that may improve life for everyone on Earth, but research often doesn't make for an eye-catching video clip.

NASA management hadn't wanted a drink experiment to distract from the key operational and research goals of our flight, and Mr. Abbey didn't want to provide more fodder for possible criticism. I understood that. I also knew that researchers paid for us to perform this experiment, and we spent valuable time training for it.

The no-win conflict between the competing objectives was my most stressful experience during the entire eight-day mission. It was even worse than leaking thrusters and space sickness. To me, achieving mission success for the scientists and engineers was more important than any other event that was out of my control.

Flying *Discovery* and Choreographing a Spacewalk

Our flight plan called for Janice Voss to use *Discovery*'s robot arm to lift the SPARTAN-204 satellite out of our payload bay and release it into space on Flight Day 5. We backed *Discovery* away from the satellite, and SPARTAN flew on its own for nearly two days, studying far-ultraviolet radiation in space with a special spectrograph.

When it was time to rendezvous with the satellite and put it back into the payload bay, Jim asked me to sit in the commander's seat and fly the rendezvous and proximity operations—a task normally flown by the commander. Over the course of one ninety-minute orbit, we would execute four midcourse correction burns, MC-1 through MC-4, to catch up with SPARTAN. Using the hand controller, I executed MC-1 and MC-2. I didn't get to fly the last two maneuvers, though. I executed the first two so precisely that we didn't need the last two. The RESIDUAL ERROR display after each burn was 0.00 in all three axes of motion—"all balls," as astronauts say. Jim remarked that he'd never seen that happen in a simulator or in flight. He actually felt guilty that I only got half the stick time he intended to give me!

When we caught up with SPARTAN, Janice grappled it with our robot arm and brought it back into the payload bay. Meanwhile, Volodya helped Mike and Bernard suit up for the spacewalk and pre-breathe oxygen in the shuttle's airlock. It would be the first time that either of them had been out in space.

NASA had planned this spacewalk to test technologies and practices considered essential for assembling the International Space Station at the end of the decade. One goal was to determine how much mass an astronaut could manipulate in the microgravity and vacuum of space. We also needed to learn whether astronauts could work outside in these space suits in temperature extremes for up to eight hours.

I was the IV (intravehicular activity) crew person for the spacewalk. My role was to read off the checklist, coordinate the activities, and monitor the overall progress from the flight deck.

Mike and Bernard both rode on the end of the robot arm to a shadowed area directly above our overhead windows. We held them there for about half an hour. We kept asking them, "On a scale of one to eight, how cold are you?" The temperature was about -150°F/ −100°C in the shade. At that point, they felt fine inside their suits.

Then Janice maneuvered them to the rear of the payload bay so that Bernard could try moving the SPARTAN satellite by hand. On the ground, SPARTAN weighed about 2,500 pounds (1,136 kg). It was weightless in orbit, but it still had mass. That is, it still took effort to overcome its inertia to put it into motion and then an opposite force to slow it and stop it. Bernard needed to avoid moving it too fast or he would lose control of it. He pushed, pulled, twisted, and turned it carefully and slowly as I called out instructions to him, "Rotate up. Rotate down. Translate left." As he moved the satellite around, he kept up a running commentary about what he was experiencing.

Now, both Mike and Bernard told us their hands and feet felt cold, and it was getting worse. Their suits weren't keeping their extremities warm.

Three hours into the spacewalk, Mike said, "My fingers feel as if they've been put in liquid nitrogen."

I told Janice, "We need to bring those guys back in." We asked Jim if he thought we should have them come back inside. He replied, "What do you think?" It was a gray area. Their lives weren't in danger, but we didn't want to put the guys into physical distress and risk them getting frostbite. Although we couldn't see the data in the ship, a sensor inside one of Mike's gloves read 20°F/-6°C. When we asked them to rate their comfort level on the scale of one to eight, both men reported "three," which we had previously agreed in training meant "unacceptably cold."

NASA almost never terminated a spacewalk before achieving all the planned objectives. We were just about to suggest to Houston that we do so now, when the CAPCOM told us that the doctors wanted Mike and Bernard to come back in. The crew and Mission Control had both reached the same conclusion at the same time.

Coming Home

The spacewalk was our last major mission objective in orbit. It was time to prepare to return home. The mission specialists stowed the experiments, and Jim and I rehearsed our reentry and landing procedures.

As we finished our landing preparations, I glanced at Janice Voss over in the corner of the mid-deck. She was casually reading a science fiction book and snacking. I asked, "Janice, how much of your food have you eaten during the mission?"

She said, "Oh, about ninety percent."

I was stunned. I think I ate at most 10 percent of the food packed for me because of my lack of appetite. I lost eight pounds during our eight-day mission. Some people obviously adjust to spaceflight better than others!

NASA learned early in the shuttle program that astronauts lose a significant volume of fluids from their bodies during a long space mission. If you didn't do something to compensate for that, you ran a significant risk of becoming light-headed or fainting when you returned to Earth's gravity. NASA was especially worried about this happening to the commander and pilot, who would be manually flying the shuttle on its final approach and landing. The space shuttle didn't have a usable auto land system. If the commander or pilot passed out during landing, the shuttle and its crew were doomed.

For this reason, NASA flight surgeons required all returning astronauts to drink a large volume of salty liquid an hour before the shuttle fired its engines for the de-orbit burn. The amount you had to consume was based on your weight, and you had to chug it all at once. I had to drink twenty-four ounces of something that was like Gatorade, only saltier. It tasted disgusting to me. I just couldn't get it all down. On my subsequent missions, I drank chicken consommé, which was slightly more palatable than salty lemonade.

When we got the word from Houston that we were "go" to land at Kennedy Space Center, we fired our orbital maneuvering system engines above the Indian Ocean and then began our inevitable fall back to Earth. We reentered the atmosphere during orbital night and were treated to a stunning display of multicolored plasma glowing outside our windows as we slammed into the upper atmosphere at 17,500 miles per hour. I was able to sneak only a few quick peeks out the windows, as I was busy monitoring our systems and preparing for the landing events that would occur in rapid succession.

As we crossed the west coast of Florida, the sun peeked above the horizon. It sank below the horizon again as we lost altitude while making the turn to line up for the approach to the runway.

We came down from the north and turned left toward Runway 15. I was surprised at how hard Jim had to pull back the control stick as he executed the banking turn. He had to fight powerful one-hundred-knot tail winds from the jet stream to make the turn. The turn pushed me down into my seat. At the time, it felt like a 5-g turn—probably exaggerated because I was experiencing gravity again for the first time in eight days.

I was grateful that Jim was expertly handling that turn, because a few years earlier, a shuttle had landed far short of its runway at the Edwards AFB lakebed under similar high-wind conditions. At Kennedy, a short landing would mean certain death.

I armed and lowered the landing gear, which locked into position seconds later as we crossed the runway threshold. Our main wheels touched down, and Jim slowly pitched us downward to put the nose wheel on the ground. I deployed the drag chute to slow us down.

"Houston, *Discovery*. Wheels stop."

I couldn't believe how heavy I felt. I took off my helmet, which seemed to weigh seventy-five pounds. I could barely lift it. We ran through our postlanding checklist and prepared to hand *Discovery* over to the KSC convoy that met us out on the runway.

I was a space veteran now. Eight days, six hours, twenty-eight minutes from launch to landing, 129 orbits of the Earth covering 4.8 million miles.

First woman to pilot a space shuttle.

A New Adventure

Just as you can't predict who will or won't get sick in space, you can't tell in advance how your body will react once you're back on Earth. Some people feel fine; others feel faint or nauseous after being in zero-g.

After a few minutes, I felt surprisingly perky and fully capable of moving around on my own. The landing team's doctors checked us out in the Crew Transport Vehicle, and then we walked out to inspect *Discovery*. When we returned to the Operations and Checkout Building a few hours after our landing, Janice and I ran up the stairs. Jim stood at the bottom, shaking his head. "I can't believe you two are running that fast! You'll be sorry tomorrow!"

He was right: delayed soreness is a universal postflight affliction.

I slept soundly that night. When I woke up the next morning, I couldn't feel anything. It wasn't like the tingling that you get when your foot is asleep. I literally couldn't feel *anything*. It scared me. I moved around to get my blood flowing, and some sensation finally started to come back slowly. The doctors told me that some people's nerve endings stop working if they haven't been used in a while. The doctors assured me I would return to normal before long. It was actually kind of fun. I felt as if I were back in zero gravity. If someone had told me in advance that this was a possible side effect of spaceflight, I wouldn't have been concerned. I made sure to add this observation to my briefing for new astronauts.

Another thing happened shortly after I returned from space: I became pregnant.

After years of bad timing throughout my career, this was perfect. Pat and I could finally start the family that we had been putting off for so long.

People thought it was odd that I became pregnant so soon after the mission. I can assure you: I wasn't pregnant in space. It didn't help that when I began giving postflight presentations, the hosts invariably introduced me saying, "Oh, and by the way, Eileen is pregnant." Everyone in the audience then apparently felt it was completely acceptable to ask me if the baby was conceived in space or why I had been so inconsiderate as to put my unborn child in jeopardy by flying when I was pregnant.

After the second time a woman asked me if I got pregnant in space, I looked her in the eye and snapped back, straight-faced, "Yes, I did. The father of the baby is an alien. They've already tested the baby, and it has green blood."

She gasped and stared at me in horror. *She believed me.* I had to apologize and tell her I was joking.

An astronaut cannot joke like that. People will believe *anything*.

You Can't Go Home Again—Yet

NASA sends a returning crew on a long series of public appearances at the end of a mission. We visit Congress, NASA Headquarters, NASA centers, and key contractor facilities and make many other side trips to thank the people who made the mission possible. In addition, NASA pays for every astronaut to return to their hometown to give postflight presentations and talk to schools and civic groups. It's a wonderful opportunity for the people you grew up with to celebrate with you and share a personal connection to the space program.

My turn came in late March, just a few weeks after I learned I was expecting. Although I was looking forward to seeing my friends in Elmira, I was already exhausted when I boarded the plane in Houston. I worried that I was putting my baby at risk, having already endured three weeks of continual travel all across the country.

We landed in Pittsburgh, where I would change planes for my flight to Elmira. The gate agent opened the door and got on the speaker. "Is Eileen Collins on this plane? Please meet with Security."

Holy cow, I thought. *What's going on?*

I met the security representative, who said only, "Follow me to the office." I feared something had happened to my husband or my family.

In the office, the manager said, "NASA has informed us there's been a threat against your life. It's considered serious enough that they don't want you to go to Elmira."

They put me on the phone to Chief Astronaut Bob Cabana. Bob told me the details behind the threat and said, "We think you need to come home, but it's your choice." It broke my heart not to be able to go to Elmira, but I wasn't about to endanger the life of this tiny baby I was carrying. I turned around and flew back to Houston.

I called my mom and dad. Of course, they were disappointed. They could also hear how exhausted I was. I told them I was done with the public affairs trips. No more speeches.

So my hometown held the big event without me. My family and my high school friends rode in the parade. I'm glad they were able to celebrate that their small-town girl could circle the Earth and help bring about the peaceful cooperation of two superpowers, or whatever stories they wanted to tell. They knew that my heart will always be in Elmira.

Back in Houston, I met with NASA Security, who eventually identified the culprit. NASA canceled my remaining appearances that year at my request. They assigned me as a CAPCOM throughout the rest of my pregnancy. Honestly, that was wonderful. I enjoyed the challenging job: it kept my skills sharp and enabled me to be at home more often. No traveling.

NASA rules for pregnant pilots were that T-38 flying was optional in the first trimester. After that, emergency ejection could cause loss of the baby and possibly my own life. So, I had a few more weeks of stick time. Thanks to those rules, I can honestly say that both of my kids have some flight hours in a T-38!

And then, for once, it felt good to be grounded.

But only temporarily.

Chapter 11

ON BOARD MIR: STS-84

A few weeks after I returned from my STS-63 mission, the astronaut class of 1995 reported for duty. Among their ranks were Susan Still, a Navy test pilot, and Pam "Pambo" Melroy, an Air Force test pilot. I was happy to see that more women were qualified to apply as pilots and that NASA had selected them. However, due to the small number of women military pilots—and the even smaller number of test pilots—the field of applicants was thin. Five years before the end of the millennium, we were the only three women pilots, and there would be no others in the entire operational history of the Space Shuttle Program stretching from 1981 to 2011.

Susan and Pam unquestionably possessed the right attitude and skills to be successful shuttle pilots. NASA had selected them entirely on their own merits. They didn't need me to pave the way.

One of the things I loved about the NASA culture was that your gender and ethnicity didn't factor into your treatment. No one discriminated against me or handled me deferentially because I was a woman. (Though people might have treated me deferentially because I was an *astronaut*.) At NASA, you earned respect based on your ability and dedication to the task at hand.

Perhaps it was a mark of that culture that Pam, Susan, and I interacted with one another throughout our careers the same way we did with our male counterparts. There was no need for a women astronaut pilot club!

Bridget

Nine months after my first spaceflight, my beautiful daughter was born. She changed my life forever, for the better. Pat and I were fortunate to have had such a healthy and happy baby. As older parents, we couldn't help but be concerned during my pregnancy.

I spent the allowed six weeks of maternity leave at home with her. I didn't suffer from full-blown postpartum depression, but I definitely had a case of the baby blues. The lack of sleep and a disrupted schedule played a large part in that. In addition, I had trouble adjusting to a slowed-down lifestyle, having lived on fast-forward for so many years. After just a few weeks, I was ready to get back to work.

Shortly after returning from maternity leave, and while sitting at my desk in the astronaut office, I found myself thinking, *I have control of my job. I have a schedule. I have a computer and a stack of papers. There are people who work for me and people who I work for. It's all logical and it all make sense! I have* control—*and things are relatively calm.*

Then I would go home to a baby who I seemingly had no control over! She would randomly cry, want to eat, and need a clean diaper. Newborn babies have a way of telling you *they* are in charge. I'm sure most new parents go through this transition and feeling of frustration. After a while, though, my daughter settled into a schedule, and I adapted to my new role.

I started thinking, *My child is actually my stress reliever.*

Before Bridget was born, I worked all day, came home, and worked all evening, seven days a week. It never let up. Now that I had a child, I decided to change my work routine to improve my efficiency and effectiveness. I would spend time with her by going for walks, reading her books, and sharing our bedtime rituals. I read technical manuals to her beginning the day I brought her home from the hospital. I'd say, "Bridget! Look! Here's a picture of the electrical circuits that go through the landing gear in the space shuttle!" It kept me in the books, and it helped Bridget to get to know my voice and eventually learn how to focus on something. (Pat also liked to read her Jack Nicklaus's book *Golf My Way*.)

I became more disciplined in my own sleep schedule, primarily to prepare for the random middle-of-the-night baby wake-up calls. I was still as productive as before, and I definitely wasn't as stressed out as when I was constantly thinking about my job. Having a child gave me a mental break and made me a better space program employee.

Training for My Next Mission

In July 1996, NASA announced my assignment as the pilot of STS-84, a mission to dock with Mir and to fly a SPACEHAB laboratory. In July 1995, STS-71 became the first space shuttle to dock with Mir, and space shuttles had visited Mir about every four months afterward. That month, American astronauts began long-duration flights aboard Mir. Our mission would take Mike Foale up to Mir and bring home Jerry Linenger after a four-month stay in space.

The projected launch date was ten months away—May 15, 1997. That date never changed during our entire training flow, unlike what usually happens. STS-84 was the only one of my four missions to launch when it was originally scheduled to.

Our orbiter for STS-84 would be *Atlantis*. Everyone at Kennedy Space Center said that *Atlantis* (engineering name OV-104, or Orbiter Vehicle 104) was "the cleanest vehicle in the fleet." Of the four active vehicles (OV-102 *Columbia*, OV-103 *Discovery*, OV-104 *Atlantis*, and OV-105 *Endeavour*), *Atlantis* historically had the fewest problems, malfunctions, surprises, and delays. That was certainly the case for our mission.

My Hairballs classmate Charlie Precourt was flying his first mission as a commander. Charlie was a whiz at languages; besides English, he spoke French and Russian fluently, and a little Spanish. That was a good thing, because our crew was probably the most internationally diverse one to date. I was amazed to hear him switch effortlessly between languages.

Our MS 1 was Jean-François Clervoy, a French astronaut representing the European Space Agency (ESA). On Jean-François's first mission a year earlier, his crew nicknamed him Billy Bob. We adopted that name, and we still call him that to this day.

Ed Lu and Carlos Noriega were the two rookies. Ed is of Chinese American heritage. Carlos was the first Peruvian-born astronaut, although he also held US citizenship.

Having rookies on our crew was a valuable experience for all the rest of us. What they lacked in spaceflight experience, they more than made up for with open-mindedness, work ethic, creativity, and thoroughness. They were also abundantly curious, asking so many questions that at times I became frustrated. Nearly every day, a training session that was supposed to end at five o'clock went on until six or six-thirty. In the past, I wouldn't have minded. Now, I had a kid at home. When they started to speak, the rest of us would half-jokingly interject, "No more questions! Questions not allowed!"

Elena Kondakova was our Russian cosmonaut crew member. She was the third Russian woman in space and the only Russian woman to fly between 1982 and 2014. I'd glimpsed her through the Mir porthole on my first flight. Now, I loved being able to work with her, both in the technical aspects of the mission as well as helping her deal with the cultural challenges of living in the United States. Her English was already pretty good and steadily improved as she trained with us. She had a great sense of humor. She laughed at our jokes and played many of her own. Her daughter, who was eleven years old when we were preparing for the mission, stayed behind in Russia when Elena came to the United States to train with our crew. Working with her proved to me once again that we have more in common with our Russian neighbors than not.

Mike Foale was my crewmate again, but his role was radically different this time. After we docked with Mir, Mike would become the fifth in the series of American astronauts to reside on the space station. We would leave him there until the next shuttle mission brought him back home five months later. Mike spent most of his preflight time in Star City outside Moscow, preparing for the experiments he would be running. We trained with him a few times for activities like Flight Day 1 setup, basic operations, and emergency egress.

While we were docked to Mir, Mike would trade places with Jerry Linenger as the resident American astronaut. Jerry launched to Mir on January 12, 1997, and we had only limited time to work with him before his flight up. When we brought Jerry home in May, he would have logged 132 consecutive days in space.

We needed to train for what might happen if Jerry was too weak to climb out of *Atlantis* following a landing emergency. At that time, we didn't know the extent to which someone who had been in space for four months might be deconditioned to gravity. Would he have a stressed heart, weak muscles, poor lung capacity, or brittle bones? Would he be able to run away from a burning shuttle? Worse yet, what if he passed out during an emergency and we couldn't revive him? We had to figure out how the rest of us—also weakened from being in space—would lift an unconscious crew member from his seat, pass him through the hatch, and lower him down to the runway, all without any assistance from ground personnel and as quickly as possible. We developed techniques and trained using a dummy. Fortunately, we never had a real emergency. We were seriously concerned about our ability to perform, although I believe we could have carried out a rescue once our adrenaline started flowing.

Our mission carried a double SPACEHAB module in the cargo bay—twice the size of the one we flew on STS-63—with a huge variety of experiments sponsored by the European Space Agency and other international partners. Jean-François was the payload commander. He, Carlos, Ed, and Elena ran most of the experiments in orbit.

Coordinating all of the various principal investigators (known as PIs, the people who designed the experiments), the training team, and the astronaut crew was a major challenge, especially when we were working across multiple languages and cultures. On one experiment, Carlos and I had to assemble unique camera equipment to record the data. It involved taking all the gear out of a locker and snapping five sections of camera lenses together. It would be a tricky process with everything floating around in zero-g. Carlos asked the PI, "Could you guys just put all this together on the ground, so we don't have to do it in space?" They looked at each other and began speaking to each other rapidly in Japanese. Then they turned back to us and said, "Yes, we can do that." That simple request saved us ten minutes and a major headache in space.

We constantly sought ways to minimize our setup time in orbit wherever possible. Carlos was in charge of powering up our laptop equipment on Flight Day 1. The need to balance electrical loads prevented us from plugging everything in before launch. Carlos had to connect the spaghetti of weightless cables once we were in orbit.

At KSC for an integration test weeks before launch, Carlos meticulously planned where all the cables needed to run. He taped them to the appropriate places on *Atlantis*'s cabin walls. Once we were in space, he discovered that someone had previously taken down the cables and put them up again—backward. Frustrated, he carried on, carefully removing the tape and reinstalling the cables, and eventually he got us back on the time line.

Star City

Our crew traveled to Russia for a ten-day training session in late October 1996 to study the layout and operation of Mir's systems. Before I arrived, I had completed one hundred hours of classroom Russian language training, split between this mission and my prior one. I learned the basics we needed to function as a crew. I thought the language itself was fascinating. I didn't enjoy the way our humorless instructor taught it—too much structure and grammar. I learned conversational Russian faster by watching *Sesame Street* on Russian TV!

We stayed in newly completed apartments at Star City. When we arrived, my Hairballs classmate Dave Wolf was already on a prolonged stay there training for his upcoming Mir flight. Our visit coincided with Halloween, and Dave planned a big party at his apartment. He posted paper invitation flyers all around Star City, with the instructions "Wear a costume!" Celebrating Halloween was not a Russian tradition, at least not at that time, so I'm sure some of the cosmonauts and instructors were intrigued and perhaps perplexed by his invitation. Several of them showed up to the party wearing business attire, which confused us. They explained to us that the Russian word *kostyum*, which sounds like the English word, actually means "business suit." It was another unexpected cultural learning opportunity.

We ate dinner one evening at Elena Kondakova's posh apartment. Elena's husband, Valery Ryumin, was the president of Energia, Russia's largest spacecraft and missile manufacturing company. We sat around a long table with fifteen to twenty people, most of whom were Ryumin's Russian friends. I sat next to Charlie, who translated the conversations for us.

In Russia, you were always toasting at every meal except breakfast. Before you ate, someone stood up and offered a toast; everyone drank, and then the toaster sat down. Then the next person stood up to offer a toast, and on and on. This went on between every course at dinner. I took small sips as the only way to survive that much alcohol. All I remember about the food was that it was delicious, and every dish seemed to contain mushrooms, a Russian staple.

Then Ryumin stood up and said something in Russian, and we all toasted him. Charlie leaned over to me and whispered, "Ryumin just toasted to himself riding on the space shuttle." We had all drunk to it without realizing what he had said. *Whoa!*

Even though he was a former cosmonaut with three previous missions, he hadn't flown since 1980. He was in his late fifties and appeared to be out of shape. As director of the Russian portion of the Shuttle-Mir program, he had the authority to assign himself to a flight in lieu of a younger cosmonaut. Ryumin said he needed to personally see and evaluate the condition of Russia's space station.

He did eventually fly in 1998 on STS-91—Charlie's final spaceflight, and the last mission of the Shuttle-Mir program. He had to lose a significant amount of weight to be able to fit—just barely—into a space suit.

Astronaut Mom

When I returned from Russia to the Houston Hobby airport, Pat met me at the jetway, holding Bridget in his arms. Then he put her down—and she walked over to me!

The first thing I thought was *I missed seeing my daughter learn to walk! What kind of mother am I?*

It took me a while to not let myself dwell on that. I couldn't change it now. I would certainly not be able to be there for everything in her childhood. I chose to be an astronaut, and there are pluses and minuses to every job. I reminded myself how lucky I was to have her as a daughter and told myself to focus on our future rather than the past.

Astronauts and flight controllers do not work a regular schedule. We put in crazy hours that revolve around the missions. For his part, Pat's airline job kept him on the road for extended periods. There were no day-care centers that could possibly accommodate our hours.

We hired a nanny. And then we had to fire her in the spring of 1997. She failed to show up one day because she was "detained" out of town—as in *involuntarily* detained, by people with badges. We contacted the agency and asked for a group of slightly more mature nannies to interview. We interviewed six women. We thought that one, about fifty years old, was perfect. When I came home from work on her first day, she said she and Bridget had gotten along well. However, she phoned me in tears that evening and said, "I'm sorry, I can't do this." She offered no explanation.

Now *I* was in tears. *Doesn't anyone want to take care of my little girl?*

Pat and I had to resolve the situation quickly. My flight was less than two months away. When Pat returned from his next trip, we opted to try my second choice, who was Pat's first choice. Stacey had been late for her interview, because she got lost in the maze of our neighborhood streets. I couldn't tolerate someone being late. If

I were delayed even five minutes for a training session because of a child-care problem, our whole crew would have to start late, and it would cause complications for our instructors and the support system. But otherwise, she seemed perfect. In addition, another astronaut had raved about how amazing she was, and Pat eagerly wanted to try her out. I agreed, with reservations.

Stacey turned out to be a gem. She ended up working for us for nine years. She was reliable, was a great teacher, and had a philosophy about discipline similar to Pat's and mine. She brought library books and art projects to our house. Bridget adored her. We kept her schedule limited to forty hours per week, because a burned-out nanny isn't good for either her or your child.

As a supplement and backup for child care, we found a good daycare center that would take Bridget on Fridays. While driving her there in the morning, I would play "The Cosmonaut's Song" on my car's cassette player so we both could practice the Russian language and accent.

On Friday mornings, I always brought the baby's sippy cups into the office or to the simulator sessions for cleaning and storing. The resulting jokes and photos kept our team laughing and lighthearted as we began our serious and grueling training sessions.

My family and I are forever grateful for Stacey. The moral of the story is that it *is* possible to be an astronaut and a mother and have an airline pilot husband. It takes a lot of juggling, a lot of energy, and a lot of love. You can find first-rate child-care professionals and know your children are well taken care of. Great nannies aren't in it for the money. They're in it because they love what they do.

Saying Goodbye

Entering quarantine, I experienced a much deeper sense of separation from my loved ones than I felt on my first launch. Pat and I had always been comfortable with being away from each other on travel, and he and I could still see each other during my quarantine, assuming he was healthy. However, an astronaut cannot be in physical contact with children in the week before a mission. Saying goodbye to Bridget was terribly difficult. I wouldn't see her for weeks. At eighteen months old, she was far too young to understand what was going on, so I'm sure it was much harder on me than it was on her.

I found myself rationalizing the situation: *After we begin the quarantine period, we might scrub, and then I could be home tomorrow!*

You come up with all these hypotheticals to help you deal with saying goodbye for a prolonged period. Military families understand this, as it is common for one spouse to deploy overseas.

It was distracting to think about what would happen to my family if something went seriously wrong on my mission. Pat said that the night after I was finally in orbit on STS-63 after so many delays, he realized how much tension had been building up in him. As a pilot, he intellectually understood the risks of spaceflight. However, he hadn't fully comprehended the emotional toll of knowing that I was going to ride on a six-million-pound controlled explosion. When he returned to his hotel room early in the morning after I launched, he opened the sliding glass door and listened to the ocean. He said, "Thank you, God," before falling asleep on the couch.

All astronauts and our families put on brave faces when talking to our friends and the media about the exciting journey ahead. We are human beings, though. Everyone, and I mean *everyone*, who watched a space shuttle launch had an unspoken fear that they could see their loved one blow up like *Challenger*.

Nobody talked about it. Everybody thought about it.

Pat told me after the mission that it actually helped to have Bridget with him for this and my subsequent launches. Taking care of someone else took his mind off his own worries.

In Orbit Again

Everything ran like clockwork in our final mission preparations. Our launch was scheduled for 4:08 a.m. (EDT) on Thursday, May 15. *Atlantis* lived up to her reputation as the cleanest vehicle in the fleet. We had absolutely no technical issues with the space shuttle throughout countdown. The only hiccup before launch was discovering that my emergency procedures checklist was missing a page on auxiliary power unit malfunctions. I noticed the missing page just as technicians began closing the hatch.

I had memorized the procedure, and I said I could go without it. Charlie was adamant: "No way!"

Someone drove a copy of the missing page out to the launchpad, and the closeout crew brought it on board. Fortunately, it didn't delay our countdown.

We took off right on schedule. It was the only time in our astronaut careers that either Charlie or I launched on time! The ascent to orbit was smooth and flawless. The upper-level winds in May were much less powerful than those during my February flight, so our ride was smoother this time. Forty-five minutes after our main engines cut off, we fired our maneuvering system engines to circularize our orbit at 185 miles above the Earth's surface.

I felt elated not to experience nausea this time. What a relief! We configured *Atlantis* for operations and settled down to sleep at eight in the morning, Houston time. We would spend our next day preparing for the rendezvous and docking with Mir.

As pilot, my primary responsibility was to care for the orbiter's systems. I had to manage the flow of cryogenic hydrogen and oxygen into our fuel cells, which combined the two elements to produce electricity and pure drinking water as a by-product of the process. When our wastewater tanks became full, I dumped the excess overboard through a port on the side of the orbiter. This produced a sparkling cloud of snowflakes that dissipated out into space. I configured and monitored our electrical system so that all of our systems and experiments had balanced and reliable power. I changed our air filters and the lithium hydroxide canisters that removed carbon dioxide from our cabin air. I even had the glamorous job of ensuring that our bathroom was clean, fully stocked, and operating properly.

True to my "Mom" call sign, I posted a chore sheet to ensure everyone on the crew equally shared bathroom cleaning duty during the mission.

I received the latest updates to our flight plan and schedules from Houston and Moscow on a continuous-feed printer that spewed out a seemingly endless list of changes. At one point, the printout was so long that the paper stretched from the shuttle's mid-deck, through the airlock and tunnel, and into the SPACEHAB module.

Flight Day 3 was our rendezvous and docking with Mir. Our final approach came during orbital night. As we slowly inched our way toward Mir's new docking module, installed late in 1995, I saw first-hand how significantly the station had changed since my last visit. Two new modules—Spektr and Priroda—jutted out from the hub, and one of the other modules had been repositioned. Mir truly lived up to the term "orbital complex," with its various antennas, solar panels, and cylindrical modules sprouting out in three dimensions.

Its many long solar panels extended out from the station in a way that minimized obstructions in the station's x axis, which was the direction along which the Russian vehicles docked. However, we were coming up from the minus-z axis—what we called an *R-bar* approach—and half of Mir's solar panels stuck down in our direction. The docking module, which was attached to the end of the Kristall module, provided the space shuttle an extra five feet of clearance below those solar panels and allowed some breathing room for our delicate final maneuvers.

Thank you, *Atlantis*, for no leaking thrusters! We established soft capture with the Mir docking module just as the sun cleared the horizon.

Visiting a Home in Space

After checking for air leaks and allowing atmospheric pressure to equalize between *Atlantis* and Mir, we opened the hatches and floated into the station. Mir's crew joyfully greeted us. Vasily Tsibli-yev was the strong and firm-minded commander (I say that because

I overheard him yelling at Russian mission control several times). Alexandr Lazutkin, whom we called Sasha, was the highly capable and somewhat quiet engineer. Perhaps happiest to see us was our colleague Jerry Linenger, who had been on board Mir since January. He was ready to return to Earth and see his wife, Kathryn, and their baby boy again. Elena brought a traditional gift of salt and bread to the Mir crew. I know she was excited to be back on Mir again.

My first impression was that this space vessel differed from *Atlantis* in so many regards. The contrasting design philosophies of Russia and the United States were clearly visible, and I could also tell that some of Mir's modules had been in space for more than ten years. Mir was warm and humid. Charlie described it as smelling like a basement in New England.* *Atlantis* was colder and drier by comparison.

Vasily and Sasha floated us through the station during their safety briefing. Mir was full of equipment, stowage bags, experiments, and other miscellaneous items floating around or stuck to all four walls along the length of every module. It was hard for more than one person at a time to pass through some areas. Ten years' worth of nonfunctional equipment that could not pass through Mir's narrow hatches was abandoned in place. The clutter of stored equipment and material struck us as a potential safety issue.

"Up" was in different directions when you floated from one module to another. You could easily become disoriented. Because of a recent fire on board the station, Vasily had put up a red fabric stripe along the wall from the service module to the docking module to lead us back to the space shuttle in case of emergency.

The ten members of the two crews gathered for a festive dinner in Mir's service module, while a cassette player entertained us with delightful Russian folk music. Each of the shuttle crew members brought a food item representing our cultural heritage. My

---

* Charlie had an interesting history with Mir: he was pilot on the first US shuttle-Mir docking in 1995, and he commanded the final shuttle-Mir mission in 1998, so he was able to compare how the station evolved—and deteriorated—in those three years. Charlie was the only American to visit Mir three times.

contribution was a box of space shuttle–shaped chocolates wrapped in silver foil, which floated about the dining area when Elena opened the box.

Vasily surprised us by producing a small glass bottle of Courvoisier Cognac. We normally sipped our beverages from squeezable drink bags, like the ones you'd give a toddler, except that ours had a pinch valve to prevent leakage. But drinking out of a glass bottle in zero-g was a totally different experience. There was no need to tilt the bottle to pour out the Cognac. It seeped up and out the neck of the bottle, adhering there by surface tension. You took a sip and passed the bottle along to the next person. After two sips each, Vasily put the bottle away for the next group dinner.

NASA's flight rules prohibited alcohol consumption on the shuttle, and we couldn't have glass bottles, either. However—we were on board Mir!

Camaraderie with friends is a huge part of Russian culture. During our visit to Star City the previous October, our crews were looking over flight plans spread out on a big conference table. After a few short minutes, Vasily said something quickly in Russian, and then he got up and walked out. Charlie translated for us: "He said, 'This is dumb. We will do this in orbit. Let's go have a cup of tea.'" We all rose, joined Vasily in another room, drank tea, and just talked.

I realized that while part of this was due to a difference in national cultures, it made sense to approach a space shuttle flight differently than a long-duration mission. Shuttle astronauts trained with intense focus and diligence, because our missions were only five to eighteen days long. We rehearsed everything; there was no time to figure something out in orbit. However, on a mission that lasted three to six months, you did indeed have time to figure things out in orbit—and maybe even discover a better way to accomplish a task.

Vasily knew it was more important for our crews to build fellowship than to review a flight plan guaranteed to change many times before the mission. So, that day in Star City, we sat and drank tea and got to know one another. We also sang Russian folk songs.

At dinner in orbit on Mir, we spoke about how different things had been between our countries just a few short years earlier. Charlie was an Air Force pilot who flew F-15 fighter planes out of Bitburg, Germany, from 1982 to 1984. At about that same time, Vasily was a pilot for the Soviet air force stationed in Eastern Europe. How could the two of them ever have imagined they would someday be sharing Cognac aboard a Russian space station?

An Aging Space Station

The station had suffered several serious incidents before our arrival. The cooling system began leaking during Jerry's stay, with so many small leaks behind the walls that they couldn't all be located and repaired. There was a near-collision with an unmanned Progress supply ship during a test of a new docking system. On another occasion, the station lost power and slowly tumbled out of control until power was restored.

The worst emergency occurred when one of Mir's Elektron oxygen generators failed. The crew had to activate supplemental chemical oxygen generators. While two of those portable generators were being switched out, one of them caught fire. It spewed flame and molten metal across the interior of the Kvant-1 module, and the station quickly filled with acrid smoke. Jerry rushed to put on an emergency respirator mask, only to discover it didn't work. Fortunately, he found another one despite the smoke and limited visibility. Some of the fire extinguishers on board were still fastened to the walls with screwed-down launch brackets. The fire cut one of the residents off from the escape route to the Soyuz "lifeboat"— the only way off the station if they had to abandon ship. It was a grave situation. Jerry later said he felt fortunate to have survived his stay on Mir.

I can only imagine what Mike was thinking. I wondered if he had second thoughts about what he had signed up for. If he had any qualms, though, he never voiced them. He was well trained and confident. Dealing with emergencies and learning to repair equipment breakdowns in orbit were some of the most valuable lessons of the Shuttle–Mir program.

On June 25, 1997, five weeks after we left Mike on Mir for his extended mission, Vasily and Sasha tested the new docking system again with another Progress resupply ship. They were unable to control the Progress during its terminal approach, and it struck a glancing blow to the Spektr module, damaging one of Spektr's solar arrays. Far worse, the collision punctured Spektr's hull. Mir began to depressurize. The crew had twenty minutes to find and isolate the air leak or else they would have to abandon the entire space station. First, Mike, Vasily, and Sasha had to sever all the cables and ductwork leading into Spektr and close the hatch to seal it off from the rest of the station. The air pressure stabilized in the rest of Mir once Spektr's hatch was closed. The crew had ten minutes of air to spare when they finished. Mike's sleep station was in Spektr, and all his personal effects were in there, as well. He lost access to his equipment, experiments, food, toothbrush, family photos, and more for the rest of his mission.

By the way, I don't blame Vasily for the collision. None of the astronauts or the cosmonauts who later tried to fly the same docking using a ground-based simulator were able to accomplish it successfully, and Vasily hadn't even been given the time to train on a simulator. The flight computer didn't accurately reflect the weight and center of gravity of the Progress. The system was also prone to data and video dropouts. Docking it was an impossible task. The result could have been much worse.

Back to our mission: one of Mir's Elektron oxygen generator systems failed briefly while we were docked. We had brought up a replacement, so we transferred it one day earlier than originally planned.

I spent much of my time filling bags with drinking water generated by *Atlantis*'s fuel cells to leave on Mir. Each of those bags would have weighed almost fifty pounds on Earth, and we sent ten of them over to Mir. We eventually transferred 7,300 pounds of equipment, spare parts, experiments, and supplies between the two spacecraft.

One of my other tasks was an extensive photographic survey of Mir's exterior using a film camera and telephoto lens. I took photos through the overhead windows on *Atlantis's* flight deck, as well as from some portholes on Mir. I looked for anything unusual I could see through the long lens, such as the several cracked or missing cells I noticed on solar panels. Those photos documented how long-term exposure to solar radiation, thermal changes, the space environment, and impacts from micrometeorites and space debris affected the materials used on Mir's exterior.

We ran a series of experiments collectively called the Mir Structural Dynamic Experiment, or MiSDE (pronounced MIZ-dee). The idea was to see how a huge structure made up of interconnecting modules responded to various forces and vibrations in weightlessness. Sensors and accelerometers throughout Mir recorded vibrations, movements, and noise. Comparing those data to Mir's design specifications would provide valuable information to the engineers developing the International Space Station.

In one MiSDE experiment, Charlie and I pulsed *Atlantis's* thrusters briefly while we were docked to measure how Mir flexed. In another test, Jerry bounced back and forth between the floor and ceiling of the Priroda module. It was easy and fun for him, but he soon started the module vibrating in a harmonic resonance that began to amplify the effect of his bouncing. Mike Foale came floating over to me in *Atlantis* and said, "Jerry's going to break apart the entire space station! Tell him to stop!" By the time I got to him, Jerry had already realized that he was creating a dangerous situation. Lesson learned: whoever designs these experiments needs to use a little common sense! I can only imagine the disastrous consequences had Priroda torqued far enough to break its docking seal with the core module.

Long Goodbyes

The ten astronauts and cosmonauts ate our final dinner together on May 21. I felt a sense of melancholy saying goodbye to Vasily and Sasha after a wonderful visit to their home in space, and I wished

Mike good luck on his upcoming adventure. I loved my brief time on Mir. I felt comfortable and at home there.

All of us seemed subdued and quiet as we prepared for our departure. Charlie and Vasily exchanged one final hug and handshake.

We began closing *Atlantis*'s hatch, and I floated up to the flight deck to run some procedures. However, I couldn't locate my crew notebook. It wasn't at my station; it wasn't anywhere on the flight deck or in the mid-deck. That notebook was absolutely priceless. It held important checklists, my camera settings for photos I needed to take, instructions for experiments, diagrams, and many personal notes. With dread, I realized that the last time I had it was in Mir's service module. If we departed, I knew I'd never see my notebook again.

I told Charlie I needed to retrieve it. Vasily, Sasha, and Mike were just about to close the hatches on the Mir side. Charlie told me to hurry over there and back as quickly as possible.

After we had just spent so much time saying goodbye, returning to Mir felt like I was imposing on them, like the guest who reappears at the host's house a half-hour after the party is over to retrieve a forgotten coat. I flew through the airlock and docking module far faster than I should have. I scraped my left thigh against the top of a panel in the service module so hard that the panel's sharp edge tore through my pants and gashed my leg. I returned to my ship and bandaged my leg. *Atlantis* remained docked with Mir, with the hatches closed, during our sleep period. But I had my crew notebook!

I'm not sure if I dreamed this, but I have a clear impression of looking at Mike through an overhead window on *Atlantis*, and him looking back at me through one of Mir's windows. We couldn't talk with each other. I gave him a small wave. I thought, *I'm not going to see him again for a long time*. He seemed quiet and reflective, not the energetic and happy fellow he usually was.

I recalled a time in the simulator for our near-Mir rendezvous on STS-63. Mike was running the rendezvous and proximity operations computer. We were discussing our need to stop at thirty feet from Mir and, per our flight rules, not get any closer. Mike suddenly blurted, "If we get to within thirty feet and we don't do anything, the gravitational attraction between Mir and *Discovery* will eventually make us touch each other." He was trying to use physics as an excuse for us to get closer to Mir. He laughed, "It's not our fault! It's Newton's law! We'll just have to explain it that way after we get back on the ground!" That was the brilliant, funny Mike Foale I knew. I was going to miss him for the rest of our mission.

We undocked shortly after awaking the next morning. Pilots on the previous Shuttle-Mir flights performed fly-arounds of the space station after undocking. Unfortunately, I didn't have the opportunity to fly that maneuver. We were testing a sensor ultimately destined for the Automated Transfer Vehicle, a European Space Agency resupply ship for the ISS. I had to back us down the minus-z axis, straight down toward the Earth. I paused for a few minutes when we were at distances of thirty meters, ninety meters, and four hundred fifty meters below Mir. The laws of orbital mechanics took over, and we gradually drifted farther away without having to fire our thrusters.

Soon, Mir was just a flashing star off in the distance.

Looking Out the Window

STS-84 was *far* more relaxing and fun for me than STS-63. After learning to perform many of my duties on my previous mission, I was more efficient now and could finish my tasks faster. I had more freedom and discretion in how I ran my day. I could actually take the time to look out the window.

I love photography, and I was able to take hundreds of photos during this mission. I challenged myself to anticipate when and where the Moon would rise and then took photos of the Earth with the Moon as a backdrop. I photographed Comet Hale-Bopp, one of the brightest comets of my lifetime. What a delight for an amateur astronomer to observe a comet from space!

My previous mission flew during February, when much of the Earth's northern hemisphere was under cloud cover, and our launch window timing caused much of our ground track to be over the oceans. On STS-84, we enjoyed better weather below us and more daylight passes over land. A laptop computer mounted on our dashboard continuously ran the World Map software program. Taking positional information from the shuttle's computers, it displayed a real-time map of exactly where we were above Earth. I still had to orient myself when I looked out the window, because north wasn't always up. World Map was a great resource to help me photograph some of my favorite places on Earth—and to learn about many surface features that I had never heard of or seen before.

Before we retired to presleep activities one night, our CAPCOM, Mario Runco, called us. He excitedly told us that we were about to pass over his hometown, New York City. I positioned myself with my face to the commander's side window and my arms outstretched, and I gazed down upon the bright lights. *Over eight million people in this one city alone*, I thought.

Then I looked up and down the Eastern Seaboard, from Canada to Florida, easily recognizing the coastline. Intense lights of all the populated areas contrasted starkly with the dark ocean. The view was astounding, something I had missed on my first flight. I couldn't tear myself away from the window. A few minutes later, Europe came into view, with its coastline also easily distinguishable against the dark seas. Cities such as London, Madrid, and Paris were immediately identifiable.

As I looked ahead to Greece, I saw the terminator, the line between night and day. It was morning in the Middle East, and I could pick out Cairo, Jerusalem, the Red Sea and Persian Gulf. My gaze turned to the curved horizon, and I saw the empty blackness of space contrasting sharply with the deep blue water and brilliant desert sands. Feeling like a flying angel, I was momentarily astonished when I realized the giant sphere below me was an oasis in the middle of nowhere.

I couldn't help but reflect on the millennia of civilizations that had sprung from the large area I was viewing all at once. I thought about the early explorers who had sailed the Mediterranean and beyond, taking risks, looking for new lands and new trade. I realized also I had just flown over billions of people, each one with their own plans, problems, and worries.

I continued watching the Earth rotate below me. We traveled over the Himalayas, India, and the myriad Indonesian islands and atolls, surrounded by gorgeous blue and turquoise waters. Finally we crossed south of the equator, and the terminator again came in to view. The sun set over Australia. Darkness again, over the vast uninhabited stretches of the South Pacific.

It was perhaps the most profoundly remarkable fifty minutes of my life as an astronaut.

What a beautiful planet. How important for all of us to take care of it!

This was an example of the "overview effect" that you often hear astronauts describe. When separated from Earth, you become even more attuned to Earth. Your love for your home planet grows, and you're filled with a desire to take care of this wonderful place. When you look the other direction and see the absolute blackness of space, you realize there are no other planets we could ever hope to reach that are even remotely like ours. It can actually be a scary thought. Astronauts come home from missions knowing we have to keep our Earth clean and livable for centuries to come.

Orbital Operations

Jerry had insisted that we bring up a treadmill on *Atlantis* for his use after we departed Mir. He wanted to keep his leg muscles in shape for his return to Earth after five months in space. Unfortunately, the treadmill was quite invasive, taking up too much space in the shuttle's small mid-deck. It was so noisy crew members had to shout at one another to be heard when someone was running on it, and we occasionally missed alarms and radio calls from Mission Control. The worst part was the vibration. I was up on the flight deck while a crew member was running on the treadmill, and the orbiter shook all around me. I worried that the excessive vibration might ruin some of our sensitive crystal growth experiments. The treadmill eventually installed on the ISS included a special vibration isolation system to prevent this issue.

The double SPACEHAB module in our payload bay actually had more than twice the volume of the single module we flew on STS-63. The extra room was a huge boost to my state of mind. Even though the shuttle's mid-deck and flight deck are roomy relative to other spacecraft, the shuttle's crew compartment still feels crowded when six or seven people are trying to do things at the same time. SPACEHAB gave us more than four times the work volume of the shuttle alone. This mission's SPACEHAB housed a suite of materials processing experiments and the European Space Agency's Biorack, a sophisticated biological laboratory sponsored by France, Germany, and the United States.

There is an endless variety of novelties and delights while living in a weightless environment. Astronauts enjoy trying experiments in zero-g that are impossible on Earth. Billy Bob and I competed on SPACEHAB to see who could make the biggest floating blob of water. Surface tension tries to hold the water together, but eventually the pulsations and airflow get to be too much, and the blob explodes into smaller globules. The crew held a contest to see who could float the farthest down the SPACEHAB tunnel without touching any walls.

—

One evening, I was in *Atlantis*'s mid-deck with Jerry and Charlie while the rest of the crew slept. Jerry gave us his detailed and blunt assessment of the safety situation on board Mir. He talked about the fire and about his experience during a spacewalk, when he came perilously close to being set adrift in space without a way to get back to Mir. As I listened to his sobering stories, I realized the risks our Mir astronauts were taking to explore space, improve technology, and promote international relations.

Things can quickly grab your attention and remind you that there is a very thin wall between yourself and instant death outside. While running some checks one morning, I noticed something odd in the commander's far left window. I found a small hole in the glass with cracks radiating out from it in a starburst pattern. I took a photo and downlinked it to Mission Control. A piece of orbital debris, probably tinier than a grain of sand, had impacted and cracked the outer thermal pane. Fortunately, there were another two panes of the window layered under the broken one holding the pressure in the ship.

One night while I was sleeping, I heard a tremendous *bang!* Was it the orbital debris hit on the window, an equipment malfunction on Mir, or something else? I never found out what caused it, which was probably even more disturbing than the noise itself.

———

As easy as I usually found it to sleep, that's not always the case with astronauts. In fact, nearly 80 percent of space shuttle astronauts needed to take a sleeping pill at least once during their missions. The sleep shifting, schedule pressure, and intensity of our workload often meant we couldn't get a full eight hours. Even if I had trouble sleeping one night, I didn't want to take any pills. I wanted to be clearheaded in case I needed to fly us home in an emergency. (It's the pilots' old "eight hours bottle-to-throttle" rule.)

The nonpilots didn't have that restriction. One of our astronauts had difficulty sleeping one evening and took an Ambien with a cup of herbal tea. The crew member later began hallucinating and talking in a semiconscious state, which alarmed the other astronauts trying to sleep nearby.

There are still only a relatively small number of humans who have flown in space. I'm sure we will continue to find unusual reactions to life in orbit that we didn't anticipate.

Back to Earth

Just about the only thing that didn't go as planned on our mission was our return to Earth. Our scheduled touchdown at Kennedy

Space Center was 7:52 a.m. on Saturday, May 24. The weather report predicted clouds over the runway at that time. However, it looked like the weather would be acceptable ninety minutes later. We were able to enjoy one bonus orbit of the Earth, although we were all eager to come home.

Everything went fine during reentry. The clouds hadn't cleared up as much as expected over the runway. When we broke through the clouds at eight thousand feet altitude, though, the landing strip was right where the flight software told us it would be.

One of my responsibilities as pilot was to lower our landing gear. On the control panel are three small windows called *talkbacks*, one for each of the two main gear and one for the nose gear. An indicator behind each window displayed one of three things: UP, DN, or a "barber pole" of black-and-white diagonal lines when the gear was in transition between the up and down position. In the simulator, when the pilot pressed the gear-down button, UP would go away, the barber pole would flash for a split second, and then the DN talkback would show for all three landing gear simultaneously. The pilot pressed the button at three hundred feet altitude, just moments before the shuttle crossed the runway threshold. With any luck, six seconds later, the gear were locked down, less than one hundred feet above the runway.

That was the way it was *supposed* to work.

As we were on our final approach, I armed the gear at three thousand feet and pushed the gear-down button at three hundred feet. All talkbacks went to the barber pole, but only one of them immediately went to DN. Another talkback hesitated and then went to DN. The third took even longer. It seemed like a lifetime, even though I'm sure it was at most a second or two. It threw me off and interrupted my callouts to Charlie. I was afraid we'd come down with a gear that wasn't locked, resulting in an unsurvivable crash landing. The strange behavior of the indicators bothered me so much that I reported it in our postflight debrief and ensured the simulator model was updated.

Just before we touched down, a strong crosswind gust caught us and blew us about twenty feet left of the runway's centerline. Charlie expertly tilted *Atlantis* so that the right tire caught the runway and stopped our drift.

Jerry Linenger surprised us all by walking down the stairs from the Crew Transport Vehicle under his own power, even after five months in orbit.

We met up with our families in a conference room at the Operations and Checkout Building, and it was so great to see Pat and Bridget again! I had to endure a few hours of medical exams and our postflight medical experiment data collections, then we left for Cocoa Beach and spent the night in a hotel. The next day, we returned to Houston and home.

Jean-François invited us all to visit France in June. Because it was only a couple of weeks after our mission, there was no way I could travel. I couldn't leave my daughter again, and Pat had his airline routes to fly. It broke my heart to turn down Billy Bob's invitation, but it was the right thing to do at the time.

Later that year, Charlie arranged for the crew and our spouses to tour Italy. Charlie had the right connections, so he planned crew speeches and diplomatic visits throughout the country. We spoke at schools, businesses, and government offices, sharing our spaceflight experiences and the importance of our mission.

The most memorable and impactful event of our trip was a personal meeting with Pope John Paul II. We arrived at the Vatican on a Wednesday, when the Pope held his reception at the Nervi Auditorium. The place was packed with people from every corner of the Earth. A huge crowd from the Pope's native Poland wore yellow and orange. The Pope addressed the crowd in many languages effortlessly and fluently. In fact, when the Pope spoke French, Jean-François whispered to me, "The Pope is speaking perfect French! He has no accent!"

We lined up to meet with the Pope after his address. We presented him with a framed set of photos of our mission and one of our space-flown crew patches. Pat asked the Pope to bless a photo of Bridget.

Being in the presence of Pope John Paul II was truly an honor and a moving experience. It reminded me that I was playing a role to bring humankind together in peace.

Chapter 12

FIRST WOMAN TO COMMAND
A SPACE MISSION: STS-93

I was eager to fly again. NASA's usual practice during the Space Shuttle Program was for pilots to fly twice from the right-hand seat before becoming eligible to command a mission. With STS-84 behind me, I had the requisite two missions under my belt. I felt ready to upgrade to commander.

I learned from the space shuttle launch manifest that the Advanced X-Ray Astrophysics Facility (AXAF) telescope would be deployed soon. AXAF was one of NASA's Great Observatories, a suite of powerful space-based telescopes that included the Hubble Space Telescope, the Compton Gamma Ray Observatory, and the upcoming Spitzer Space Telescope. I thought flying AXAF into orbit would be a fantastic way to contribute to the science of astronomy.

Soon after STS-84, I ran into Bob Cabana, chief of the Astronaut Office, in the hallway. While chatting with him, I felt compelled to put my hat in the ring. I said, "I would love to fly that AXAF mission. I don't want to ask for a flight, but I've always loved astronomy, and I'd like to get to know more of the professional astronomy community. I think flying AXAF would be a great way to be part of that."

Bob said he'd look into it.

I'd shared an office with Bob during my ASCAN days, and I considered him a friend. However, I later regretted impudently asking my boss to assign me to a specific mission; it felt too brazen. I never talked to him about it again after making my request. But Bob was true to his word.

Several months later, Jim Wetherbee (now deputy director of JSC) called me to his office on the eighth floor of Johnson's Administration Building. He told me I was assigned to the flight, and that JSC Director George Abbey wanted to speak with me. I sat at Mr. Abbey's conference table in his office on the top floor and gazed out the windows at the beautiful view, spanning Galveston Bay to the Houston skyline. Both Jim and Mr. Abbey were gracious and excited when they gave me the formal news! We discussed the AXAF mission and my role as the first woman to command a US space mission.

Mr. Abbey added, "The First Lady wants to announce the flight. They want to make a public event of it. You're going to the White House—next week."

The Big Announcement

The suddenness of all of this took me aback. I didn't want to be the center of a media circus. NASA Headquarters made it clear that this was a must-do, though. I didn't have a choice. They asked me to write some short remarks for the occasion.

Pat and I drove to San Antonio that weekend. I wrote my speech on the three-hour drive back home while trying to keep two-year-old Bridget entertained in her car seat. On Monday, Eileen Hawley (astronaut Steve Hawley's wife) from NASA's Public Affairs Office helped me polish the speech and crystallize my key takeaway message: dreams *can* come true.

The big day dawned on Thursday, March 5, 1998. Pat, my cousin Mary Kay Morin, and I arrived at the White House. As we walked down the hallway of the West Wing toward the Oval Office, I was astonished to see Dr. Sally Ride—the first American woman in space—step out of the Roosevelt Room to greet me. It was the only time I ever met her in person.

My family and I met Bill and Hillary Clinton in the Oval Office. I could immediately see why women found him so charismatic and charming. He spoke of meeting and being impressed by the famous physicist Stephen Hawking. Mrs. Clinton talked about wanting to be an astronaut when she was a teenager. President Clinton was interested that Pat was a scratch golfer. He even mentioned Pat's golf game in his remarks to the press!

Someone came in and said, "It's time." I walked across the reception area to the Roosevelt Room and peeked in the door. The place was packed. Cameras, media people, and bright lights lined the entire back wall of the room. I saw Sam Donaldson and many other reporters I recognized from the national news. Sally Ride was in the front row.

For perhaps only the second time in my life, I had a full-on feeling of panic. Over the course of three or four seconds, cold sweat broke out over my body, starting at my forehead and face and working its way down. Pat was behind me. I stopped in my tracks, turned to him, and said, "I'm not going in there." I meant it, too.

I had to play a mind game with myself. I looked down the hallway, then to the left and to the right. I said to myself, *Wait a minute. You're not going in there as Eileen Collins. You're going in there as the first woman commander. The commander can do this.* And then I was okay.

Bill Clinton caught up with us at that moment. The *president* and the *space shuttle commander* walked into the Roosevelt Room together.

Mrs. Clinton introduced me by announcing that as commander of STS-93, I would take "one big step for women and one giant leap for humanity." NASA Administrator Dan Goldin added that "To discover new worlds, we must break down old barriers."

Then it was my turn. I spoke about my childhood dream of one day being a pilot or an astronaut. I said I'd always been fascinated by astronomy and science. I thanked the women barnstormers, the women air service pilots (WASPs) of World War II, the Mercury 13, the first women military pilots, and the first women astronauts as my inspirations and role models. I thanked my parents, my family and friends, and especially Pat.

I added, "Throughout my life, and in particular in my career in the US military and with NASA, I've been given important jobs and responsibilities, and I now accept this responsibility with all the determination and the motivation and the diligence that I've had in all the other challenges I've faced." I vowed that my crew and I would focus on the mission 100 percent, and that we would make it one of NASA's greatest successes.

I ended with "It's my hope that all children—boys and girls— will see this mission and be inspired to reach for their dreams, too, because dreams do come true."

President Clinton concluded with this aspirational statement:

> *The greatest mark Colonel Collins will make will not be written in the stars, but here on Earth, in the mind of every young girl with a knack for numbers, a gift for science, and a fearless spirit. Let us work to make sure that for every girl and for every boy, dreams and ambitions can be realized, and even the sky is no longer the limit.*

The Crew

I wasn't the first person named to our crew. That was Michel Tognini, representing the French National Centre for Space Studies (CNES). Administrator Goldin announced Michel's appointment during a trip to Europe in December 1997. Michel was the veteran of a 1992 mission to Mir with a Russian crew. Since 1995, he had been working in Houston in operations planning and robotics. He was assigned as MS 3.

Jeff Ashby was our pilot, taking his first spaceflight. My Air Force friend Cady Coleman was our MS 1. She and Michel would be responsible for deploying the AXAF telescope and then conducting experiments in orbit during the remainder of our mission.

Winston Scott was the MS 2 on the crew list that went up the chain for approval. When the official roster came back, Winston had been replaced by Steve Hawley. NASA Headquarters insisted that our crew include a professional astronomer and someone who was familiar with the Inertial Upper Stage (IUS) rocket that would insert the telescope into its final orbit. Steve Hawley was the only available astronaut who met both of those requirements.

Winston had heard about his recommendation for the mission, and of course he felt bitterly disappointed to learn later that he wouldn't be flying after all. I commiserated with him. I told him how the same thing happened to me before my first flight. That's still no consolation when you anticipated taking part in such an interesting mission, only to find out that management replaced you.

Steve Hawley was in the first class of space shuttle astronauts selected in 1978, and he flew two satellite deployment missions before the *Challenger* accident. He was on the crew that deployed the Hubble Space Telescope in 1990. Steve left the astronaut corps and served as associate director of Ames Research Center for two years. He returned to JSC in 1992 as the deputy director of Flight Crew Operations. Then NASA activated him to flight status in 1996 for the second servicing mission to the Hubble Space Telescope. He was unquestionably one of the most experienced and knowledgeable mission specialists in the Astronaut Office.

Steve had pushed back on the assignment because of his current role as deputy of Flight Crew Operations. He felt it was inappropriate for managers to take the place of astronauts who needed the flight experience. However, Mr. Abbey persuaded him he was critical to the mission.

I was quite frankly worried that since he was so experienced, Steve might come in and try to run the show, telling us how to do our jobs. I was totally wrong. He was a fantastic crew member. He knew when to speak up and when not to. He asked questions to help us think through difficult problems rather than being directive. Steve provided wisdom and guidance to each of my crew, including myself. He linked stories from the early shuttle days to our unique issues with AXAF.

Schedule Compression

With our crew assigned, we had less than nine months to prepare for an ambitious mission. There was no time to spare.

I couldn't let attention from the public and the media interfere with us or distract us from our preparation. I told Public Affairs that

I'd spend one day giving interviews at Johnson Space Center, and that was it. My guiding philosophy throughout my entire Air Force career was "Stay focused on the mission." If I was paving the way for future women commanders at NASA, the last thing I wanted to do was make a mistake because I wasn't completely engaged in our training.

My crew needed to be fully committed to the mission, and as commander, I had to set the example. We needed to be perfect on this flight. I was delighted that NASA supported my request to limit media events.

Next, I sat down with lead instructor Lisa Reed and training manager Stacie Hughes to create a matrix of all of the flight's scheduled activities. We assigned every activity, from running experiments to changing the cabin air filters to cleaning the galley, to both a primary and secondary crew member. This process quickly and clearly showed whether particular days were overscheduled or if a certain crew member was overloaded with work.

Once we balanced the workload, we had to determine how long it would take to train the crew. The rule of thumb is that astronauts spend about eight hours training on the ground for every hour required to perform a task in space. Lisa and I realized we couldn't possibly train for all of the scheduled activities in the limited time available. We started tossing out low-priority tasks or looking for ways to provide refresher training rather than a complete course on some systems or procedures.

Our Payload

The AXAF is unlike any other telescope. It's a long, narrow, tapered cylinder with an octagonal box at the wide end and is completely wrapped in silver mylar insulation. Once it was flying free in space, it unfolded two wings of solar panels protruding nearly thirty feet from each side of the spacecraft module that contains the instrumentation, computers, communications systems, and thrusters to help point the telescope.

The telescope needed to be in a high orbit, far above the Earth's damaging radiation belts that could interfere with its observations. AXAF's orbit was a long ellipse that was 8,900 miles above Earth at its low end and 83,600 miles at its farthest point—one-third of the distance to the Moon.

The space shuttle couldn't fly that high. Boosting AXAF to its operational orbit was the task of the Inertial Upper Stage (IUS), a two-stage solid-fuel rocket temporarily attached to AXAF's space-craft module. AXAF and the IUS together were fifty-five feet long, merely inches shorter than the shuttle's cargo bay.

At 12,930 pounds for the AXAF telescope and 30,582 pounds for the IUS rocket, ours was the heaviest payload that a space shuttle ever carried into orbit. To save weight, NASA stripped down our orbiter *Columbia* wherever possible. Two of the five pairs of tanks holding cryogenic oxygen and hydrogen for our fuel cells were removed, which meant that we could fly only a relatively short mission. Technicians pulled more than one thousand pounds of ballast from the rear of the ship. Our crew comprised only five people instead of the usual six or seven. We would be in orbit only about five days, which reduced the quantity of consumables we needed to carry. Technicians removed *Columbia*'s robot arm from the payload bay.

Cady and Michel practiced the procedures for checking out our precious payload and deploying it into space. Unlike the Hubble Space Telescope—designed for servicing and upgrades in orbit—there would be no second chances if anything went wrong with AXAF or the IUS. Once we released the payload from the space shuttle, that was it. If something went wrong after deployment, AXAF was a dead loss of $1.6 billion worth of hardware. We had no capability to recapture it and bring it back on board *Columbia*, and no future mission could repair it in orbit.

Cady and Michel needed to be 100 percent certain that the pay-load was operating flawlessly before they deployed it. They trained for contingency spacewalks to deal with situations like not getting full motion from the tilt table, which rotated the telescope up and out of the payload bay to avoid hitting the shuttle during deployment. If we couldn't resolve a technical problem while the IUS and AXAF were still in the payload bay, we would have to bring the whole package back home.

Michel's presence on our crew led to an awkward situation during our training. Because the IUS could also be used to launch national security payloads, federal ITAR regulations prohibited foreign nationals from seeing the Boeing plant where it was manufactured or being briefed on the details of IUS technology. As commander, I spent many hours negotiating how Michel could access the information he needed to fulfill his role on the mission. But, despite my efforts, we ended up being the only space shuttle crew to fly an IUS without training at the Boeing plant.

New Commander, New Pilot

My pilot, Jeff Ashby, and I were both new in our respective roles. He and I practiced together closely to perfect our cockpit resource management techniques as well as all of the technical challenges of flying the space shuttle and operating its systems.

The Mission Specialist 2 on the shuttle's crew serves as the flight engineer and has important roles during ascent and entry. Steve flew as MS 2 on all four of his previous missions. He knew his job thoroughly. Jeff and I appreciated having Steve's expertise on the flight deck.

AXAF's mass, and its effect on *Columbia's* center of gravity, led to some sobering challenges during our simulations of engine-out abort scenarios. Every shuttle mission had a unique launch profile based on its weight, payload, launch inclination, and other factors. If you graphed your altitude against your airspeed, you saw situations where losing one or more engines during ascent meant that the shuttle simply couldn't go fast enough to fly to a safe landing place or give the crew enough time to bail out. Those were the *black zones* in the launch profile: loss of vehicle and crew. One example was if two of the shuttle's main engines shut down while the solid rocket boosters were still firing. Green zones were survivable. Yellow zones were *maybes*—maybe some of the crew could escape to safety, but probably not all of them.

Our black zones lasted longer than on a typical mission, and there were more of them, due to our heavy weight and aft center of gravity.

Even if we did make it back to KSC in an abort, it wasn't certain that we could land safely. The far-aft center of gravity increased the likelihood of dangerous pilot-induced oscillations during landing. These were situations where the commander's attempts to control the shuttle would result in overcorrecting in the opposite direction.

Nobody had ever attempted to land a space shuttle with such a heavy payload on board. The structural engineers weren't entirely sure the landing gear could handle the extra weight if the commander landed at a greater than normal descent rate.

This was *not* going to be a routine mission, and we hoped *Columbia* would behave herself during our ride to orbit.

Delays

As expected, NASA pushed the launch date back several times. First, NASA moved our flight to early January, then to late January. Then, the telescope's delivery date to Florida moved to late January, meaning we couldn't fly until March.

We were happy to have the breathing room and extra training time afforded by these initial delays. I checked in constantly with the crew, asking, "Do you need more training time than you're getting on this? Do you need less? You tell me, and we'll get it fixed." I wanted to make sure we had an open feedback loop—that they could tell me anything, and I would do my best to address their concerns.

After a naming contest open to schoolchildren nationwide, NASA announced in December that AXAF was now known as the *Chandra X-Ray Observatory*. Its namesake was Dr. Subrahmanyan Chandrasekhar, a brilliant Indian-born astrophysicist who spent his career at the University of Chicago and who won the Nobel Prize for physics in 1983.

Right after that December announcement, NASA pushed our launch to April due to another delay in telescope testing. Chandra finally arrived at KSC in early February. At the same time, NASA moved a space station assembly flight ahead of ours on the launch manifest. This meant we would launch in early July.

Then an IUS failed on an Air Force Titan IV launch in April 1999, leaving its satellite stranded in a useless orbit. NASA couldn't mate Chandra to its IUS until the Air Force and Boeing investigated the failure. They eventually traced the problem and found that our IUS didn't have the same configuration. However, Chandra failed a test at KSC when its sunshade wouldn't open. Thank goodness we found and corrected that problem on the ground!

Pushing our launch to July boxed us into a scheduling corner. *Columbia* was due for a fifteen-month-long orbiter maintenance downtime period at Palmdale, California, after we returned from our flight. NASA was going to completely replace *Columbia's* computers and cockpit displays and upgrade many other systems on the vehicle. Orbiter maintenance downtimes were tightly scheduled; they required a highly skilled workforce, and NASA could have only one orbiter at Palmdale at a time.

Because *Columbia's* airlock was inside the crew compartment and not the payload bay, it was the only orbiter with a payload bay long enough to accommodate Chandra and the IUS. If our launch date slipped any further, we'd run into *Columbia's* maintenance period. Chandra couldn't fly if *Columbia* couldn't fly. That would delay our mission by at least fifteen months.

These scheduling variables may seem frustrating, distracting, and even overwhelming with unknowns hitting from every direction, but NASA routinely deals with multiple conflicting factors in all its operations. Our professionals calmly work through the problems and confidently choose the best option. It's one of the things NASA does best.

My mantra was not to worry about things that were beyond my control. As commander, though, I was seriously concerned. I started thinking about options to keep my crew—and myself—motivated, engaged, and sharp if we encountered such a long delay.

Final Simulations

Our crew enjoyed working with Lisa Reed and her training team. We spent so much time together in mission simulations that we felt like an extended family.

When Lisa purchased a new car, we took her out to a local hotel restaurant after work to celebrate. We urged her to talk about how excited she was about her car, and then we asked her to show it to us. She was shocked to see that it wasn't where she'd parked it. We waited a few beats, and then one of us admitted to secretly pilfering her car keys from her purse and moving the car. That was a typical practical joke for NASA teams who spent many long hours together.

Speaking of cars, I renewed my vehicle registration on my 1994 Camry, and Texas issued me license plate number MRS-93R. Everyone thought I ordered a vanity plate, but it was strictly a coincidence. Nonetheless, the training team enjoyed ribbing me about being "Mrs. 93R." What did the R stand for, though?

As our crew proficiency grew, we brought in Flight Director Bryan Austin and his Mission Control team for integrated simulations. Those sims were practice runs for both the flight controllers and our crew, helping us troubleshoot problems and communicate effectively in high-stress malfunction situations. Then we brought in the teams from Boeing, TRW (who built Chandra), the Smithsonian Astrophysical Observatory in Cambridge, Massachusetts (who operated Chandra), and the Air Force IUS Operations Control Center at Sunnyvale, California, for joint integrated simulations with Mission Control and the crew.

Simulations were our lifeblood. The best way to understand how a complex system works is by studying the ways in which it fails. Lisa and her team put together an endless number of highly realistic sims that left us feeling prepared for nearly any contingency.

It can sometimes feel like the sim team is trying to punish you with malfunction after failure after warning, often several times a minute. The shuttle is unlikely to experience that many failures in such a short period of time. The sims do prepare you, though, for the sensation of urgency and the need for instant, informed analysis and response when a failure occurs in real life. From having practiced together so many times, the crew learns how to deal with crisis almost instinctively. You've trained out panic and emotional responses as much as humanly possible. You and your crew "die" many times in the simulator, but at least you can walk away and try again.

Traditionally, in the final sim before the crew goes into quarantine, the training team will run a nominal scenario in which the crew can achieve orbit safely. It's considered bad luck to have the final simulation be one in which the crew isn't successful. The crew should complete their training with confidence, knowing they are in top shape for the mission.

That said, the training team won't make it *too* easy.

In our final simulated ascent, Lisa and her team caused a failure in one of our three AC power distribution systems, which resulted in numerous electrical component breakdowns. They also tossed in a small fuel leak in an engine, which led to an early shutdown and a lower altitude than planned. (Keep this combination of unique failures in mind as you read on.) Our crew handled the situation and made it into orbit successfully.

We were ready to go.

We conducted the traditional preflight press conference at JSC on the afternoon of July 7. As I expected, reporters were more interested in the first female commander than they were in the priceless scientific instrument we were carrying into orbit.

A reported asked me, "What does your daughter, Bridget, think about her mommy flying the space shuttle?"

I answered, "Bridget thinks *everybody's* mommy flies the space shuttle." Although the room erupted in laughter, it was actually true. Bridget was only three and a half years old, and her playmates were the children of astronauts. I was just like all the other parents in her world!

Quarantine

Our crew entered quarantine a week ahead of our scheduled launch, which was set for July 20, 1999. The quarantine facility at Johnson was an older one-story building toward the back of the Center. It was the same building where I had initially interviewed for NASA in 1989 and was similar to temporary quarters at a military base, with bedrooms, a kitchen and dining area, workspace, and two large conference rooms.

Our launch day schedule had us waking at about seven-thirty in the evening, because our launch was scheduled for 12:26 a.m. (EDT). Therefore, we needed to shift our sleep cycle to acclimate us to being awake in the middle of the night. To effect a thirteen-hour change in our sleeping and waking cycle, we preferred that our sleep shift become a "slam shift," where we stayed up an extra seven or eight hours the first night.

After we arrived at the crew quarters and dumped our bags in our bedrooms, we stayed under bright lights in the conference room until perhaps three in the morning. Then we dimmed the lights and stayed awake another two hours or so. We slept for eight hours, and then we stayed under the bright lights again for a few extra hours longer than normal. After a few days of sleep shifting, our body clocks lined up with the mission schedule.

We were slightly groggy the first couple of days of quarantine. NASA tried diligently to minimize any outside distractions. We took care of last-minute business, prepared ourselves for our mission, or chilled out watching movies or sports on the TV.

One of the staff sought me out early in the quarantine: "Eileen, you've got a call from Madeleine Albright."

I must have been bleary from sleep shifting. "That's kind of funny," I said, "the secretary of state has the same name."

"Eileen, it *is* the secretary of state!"

I'm not sure how Secretary Albright got a call through to me, but she wanted to offer her encouragement. She said, "Good luck! We're cheering for you!"

—

Our crew flew to Kennedy Space Center in T-38s and landed just after sunrise on Friday, July 16, 1999. In retrospect, that was probably the last *normal* event of the next crazy week.

We checked into the crew quarters at the O&C Building and went to bed. In the predawn hours of the next morning, the news reported that a small plane carrying John F. Kennedy, Jr., his wife, and her sister had disappeared somewhere between New Jersey and Martha's Vineyard. Like the rest of the country, we were shocked and saddened to watch the terrible story unfold over the weekend.

Despite this awful news, excitement was building for our launch early Tuesday morning. July 20 was also the thirtieth anniversary of the Apollo 11 lunar landing. Neil Armstrong, Buzz Aldrin, and Michael Collins were coming to KSC for a celebration of that anniversary, and they planned to stay for our launch. I invited Lisa Reed and her training team to KSC as our guests to see us off. They were so wonderful in helping us prepare that I wanted to be sure they had a small reward for their hard work.

The US women's national soccer team had just won the FIFA Women's World Cup. Mrs. Clinton and the team were coming to KSC to watch our launch.

The First Lady's office called to ask if she could visit us at the crew quarters. I certainly appreciated her enthusiastic support. However, I realized that inviting a public figure to see the crew during quarantine would set an extraordinarily bad precedent. I also feared it would look like I was receiving special treatment. Besides, we needed to spend our energy and time preparing ourselves for the mission. I politely declined the request. Mrs. Clinton graciously accepted my decision.

And then there was the escalation of a long-standing *gotcha* war between Cady Coleman and Rick Linnehan, our astronaut support person for this mission.

Cady and Rick were members of the astronaut class of 1992, and both were true practical jokers at heart. At some point during their

ASCAN training, Cady hung up an autographed calendar of *Home Improvement*'s "Tool Time Girl" in Rick's shower. He retaliated by hanging a full-length poster of the actor and fashion model Fabio in her shower.

Prank wars between astronauts could escalate for years. And so, seven years later, in quarantine at the Operations and Checkout Building, Rick asked me if Fabio could attend the astronaut beach house barbeque before the launch. Rick had already invited Fabio and his entourage to KSC as his personal launch guests, "on behalf of Fabio's admirer"—Cady. Fabio was understandably flattered, and he kindly sent flowers to Cady at the crew quarters.

I sarcastically said, "Rick, let me think about this for a minute. *No.* No, you cannot have Fabio at the beach house. We're in quarantine. It's not a good idea. I already said no to the First Lady. I'm not going to say yes to Fabio."

He replied, "Okay, I understand."

Later in the day, Rick asked me, "Can Fabio come to the crew walkout and wave at Cady from the front row?"

Again, I told him it wasn't a good idea. I didn't want her to be startled by seeing Fabio just a few hours before she deployed a billion-dollar telescope.

"How about if Fabio just calls her on the phone while we're at the beach house?" he asked later.

He was just not going to give up, so I relented. I said, "That's okay. We can do that."

And that is why Cady received a call from Fabio at the beach house barbeque the night before our first launch attempt.

After our mission, Jim Wetherbee chewed out the entire Astronaut Office, I assume because much of this story made the front page of the *Florida Today* newspaper. He said, "If you're going to play practical jokes, there are two rules. First of all, you have to be sure all recipients will be receptive to the joke. Second, you have to understand how it will be perceived if it becomes public."

Launch Surprise

Our crew awoke in the early evening of July 19, ate our breakfast, and suited up. As we exited the elevator at the ground floor of the O&C Building, the training team cheered us from a roped-off area next to the door. We exchanged waves and smiles.

As commander, I was the first to take my seat in *Columbia's* cockpit. Jeff, Cady, and Steve joined me on the flight deck. Michel had the mid-deck to himself. He and Cady would switch positions when we came back home.

Everything looked good during the countdown. At T minus ten seconds, the computer commanded, "Go for main engine start," which triggered the sparklers under our engines to burn off any stray hydrogen gas in the area before the engines lit.

T minus nine, eight . . .

"Cutoff! Give cutoff!" came an unexpected call over the communications loop. The countdown halted at T minus seven seconds, a half-second before the main engines were supposed to ignite.

We were not going to space tonight.

A large countdown clock in the cockpit—which wasn't taking data from our flight computer—counted down to zero and then began counting up. Obviously, we ignored it. Confusingly, the data feed on the CRT in front of me counted down to seven seconds and then became stuck in a strange cycle. It kept alternating, "8, 7, 8, 7, 8, 7 . . ." I had never seen anything like that in the simulator. It distracted me for a few critical seconds.

I snapped back into the moment. *What caused the cutoff? Were we going to stay put? Or were we going to have to perform a Mode 1 emergency egress?*

If there was something terribly wrong on the launchpad, the White Room would swing back to the shuttle, we'd pop the hatch, run across the launch tower, and ride the emergency slide wire baskets to the blast bunker at the perimeter of the launch area.

I listened to the chatter on the radio and told the crew to sit tight for a moment until we received directions from the Firing Room.

You instinctively want to unstrap yourself to be ready to get out of there fast if you need to, although the official procedure is to stay strapped in and wait for instructions.

We eventually learned that a sensor had indicated a high hydrogen gas concentration in our engine compartment, which could cause an explosion. The fire and rescue team and closeout crew were put on alert to come get us if necessary. We ran through our checklist to recycle back to a safe configuration.

I never felt like we were in danger out there. The situation would have been a lot more serious had the engines started and then shut down. The closeout crew arrived, and we unstrapped and went back to the crew quarters.

It turned out that engineer Ozzie Fish in the Firing Room saw the high hydrogen reading and made the gutsy call for a manual cutoff rather than waiting for the automated ground launch sequencer to pick it up. Engineers eventually traced the reading to a malfunctioning sensor. Fish had no way of knowing at the time that it was bad. NASA later commended Fish for his quick thinking and courage in calling to stop the countdown before the engines fired.

Procedures required two days for the technicians to replace the sparklers on the launchpad before we could try again. Had our engines ignited and then shut down, flight rules required swapping out the engines—a process lasting at least a month. That would have taken us into the maintenance downtime.

A cutoff one second later might have delayed our mission by more than a year.

Second Attempt

We were back at the launchpad two nights later. In the White Room, the closeout crew member who helped me into my parachute harness gave it a sharp yank to cinch it tight. It wrenched my back, and I instantly felt a searing pain down my right side. I was not about to say anything about it, though. I would not permit my discomfort to delay the mission.

Rick Linnehan entered the crew compartment to help me strap in. He could see I was in pain. I told him what had happened with the harness. He reached into his pocket, took out three pills, and told me to take them immediately.

"Rick, you're a veterinarian, not a doctor!" I said.

"Trust me, Eileen. Take these."

I wasn't about to take any pain pills before a launch! Rick was quite insistent. I put them into my pressure suit pocket for later. The pain was excruciating. I was gutting it out in my seat.

We were about twenty minutes from launch when I saw a flash of lightning outside the cockpit windows. Our forecaster had previously given us a 100 percent likelihood of favorable weather. Lightning was not a good sign. Flight rules did not permit a launch if lightning was within ten miles of the pad or our flight path within thirty minutes of T-zero. We had a forty-seven-minute launch window, so we stayed strapped into our seats and waited for the storm to clear.

It began raining just offshore. The winds kept the storm away from the launchpad. As we neared the end of our launch window, the ops manager on the Mission Management Team asked if it would be possible to extend the launch window by ten minutes. With that delay, we would lose one of our two backup opportunities to deploy Chandra if something didn't go right on our first attempt. We agreed that it would be worth the risk if it meant we could launch tonight.

The storm still didn't clear. We could add another seven minutes to our launch window if we sacrificed our one remaining backup deployment opportunity. While management debated that option, though, the thunderstorm moved onshore. We were not going anywhere tonight, except back to the crew quarters. We'd recycle and come back out again tomorrow night, if the weather cooperated.

I swallowed Rick's pills when I returned to the crew quarters. Although they relieved the pain a bit, I required physical therapy for my back.

I was glad not to carry that pain into orbit with me.

It was about this time that Lisa Reed and her team realized what the R stood for in my MRS-93R license plate: "R for 'Recycle'!"

The Third Time Is Charmed

The next night would be our third attempt at launch. Would we *finally* be able to blast off? Everyone felt a little anxious and a little silly. Superstitions and rituals were a fun way for us to break the tension. Who or what was jinxing the mission? Would we have to sneak one of our crew on board in disguise to break the curse?

We suited up and piled into the elevator in the O&C Building. When the doors opened at the ground floor, our crew pressed up against the side walls of the elevator, hiding from everyone out in the hall. I held a crew photo out through the door and then cautiously peeked my head out. Our training team was standing where they were for the previous two walkouts, but this time they all held up hand-drawn paper masks with cutout eyeholes.

When we got to the White Room, someone gave Steve Hawley a paper bag to put over his head so that *Columbia* wouldn't recognize him. Steve had sneaked onto *Columbia* wearing a Groucho Marx disguise on his mission in 1986 after four unsuccessful launch attempts.

Midnight passed, and now we were into the early morning of July 23, 1999. The countdown proceeded smoothly, interrupted for seven minutes because of a communications problem with a tracking station. The weather was great, and our ship was behaving herself. With the comm systems problem fixed, we were now scheduled for launch at 12:31 a.m.

The main engines ignited at T minus 6.6 seconds.

A gold pin used to plug an oxidizer post in our right engine shot out of its hole at the speed of a rifle bullet and impacted the engine bell. Each engine nozzle is lined with small tubes carrying liquid hydrogen, which cool the engine bell and keep it from burning through in the intense heat. The pin hit and ruptured three of those tubes. They began leaking hydrogen. Had two additional adjacent tubes ruptured, the engine would have burned through and exploded sometime during our flight. Had the engine's oxidizer post failed because of the ejected pin, the engine would have exploded.

No one was aware this had happened. There were no warning lights in the cockpit or in Mission Control. We launched with a compromised engine.

At T-zero, the solid rocket boosters fired, the eight hold-down bolts simultaneously severed, and we blasted off.

An alert tone sounded in my headset five seconds after we lifted off. "FUEL CELL PH1" was the indication on the control panel. The "H2O LOOP" light flashed momentarily on the caution and warning panel.

Something was wrong with our electrical system.

We cleared the launch tower. *Columbia* began its preprogrammed roll to put us into a heads-down orientation. Normally, the commander calls to the ground to acknowledge that the shuttle has begun its roll program. I called the warning light to Mission Control. "Houston, *Columbia* is in the roll, and we have a fuel cell pH number one."

At that same instant, the flight controllers in Houston saw the primary digital control unit (DCU A) on our center engine and DCU B on the right engine drop offline. Our onboard displays were not programmed to show us that information. There were two of these DCUs on each engine, a primary and a backup. If both controllers failed on one engine, the computer would shut down that engine, and we'd have an immediate abort situation.

Houston didn't know if it was a real issue or an instrumentation problem. They immediately instructed to us to disable our AC bus sensors. That would keep the remaining DCUs working in case of additional intermittent power problems. Houston told us that it appeared we'd had a momentary short on our AC1 electrical system.

To top it off, the booster officer in Mission Control saw a (false) reading that the right-hand solid rocket booster was just about to lose hydraulic pressure, meaning that it couldn't be steered. Again, it was probably fortunate to those of us on board *Columbia* that we didn't know what they were seeing on the ground!

This all happened within the first thirty seconds of our flight!

Our engines throttled down as we entered the period of maximum dynamic pressure, Max-Q. Just as we started throttling up, CAPCOM Scooter Altman called to us, "*Columbia*, Houston, we are critical to AC2 on the center engine, AC3 on the right. We lost DCU A on the center and DCU B on the right."

Jeff acknowledged, "Copy that."

Scooter said, "And *Columbia*, Houston, you are go at throttle up."

I replied, "*Columbia*, go at throttle up."

The fuel leak from the broken cooling tubes began affecting the performance of the right engine. With fuel and oxidizer temperatures higher than expected but with lower thrust than required, the system started feeding more liquid oxygen to the engine to make up for the shortfall in hydrogen.

Seconds ticked by. Two and one-half minutes after launch, the solid rocket boosters burned out and separated. Scooter called up, "Performance nominal."

A minute later, we received the call that we could make it to the trans-Atlantic abort site at Ben Guerir on two engines if necessary.

The booster officer informed the flight director that it appeared there was a fuel leak in the right engine and that he was monitoring the situation.

We climbed ever higher and faster. We could now make it to Banjul in the Gambia on one engine if needed. Then we got the blessed callout I'd been waiting for, "Press to MECO."

You could hear the excitement in my voice when I replied, "*Columbia*, press to MECO!" That meant we could now make it safely to orbit if one of the engines failed.

As we neared the "fine count" for main engine shutdown, just over eight minutes after liftoff, I once again watched the little triangular "bug" crawl across the time line on my computer display. Before the bug hit the MECO position on the display, the engines unexpectedly shut down early. We did not know at the time that we ran out of oxidizer, rather than the engines shutting down via computer command.

A few seconds later, our external fuel tank separated as programmed. We were in space.

Scooter called up, "*Columbia*, Houston, we see a fifteen foot per second underspeed. OMS 1 not required." Although we were about ten miles per hour short of our intended velocity, we would not have to use our orbital maneuvering system to make up for the shortfall in speed.

I acknowledged, "Copy, OMS 1 not required, and it's great to be back in zero-g again!"

On the internal comm loop in Mission Control (unheard by us on *Columbia*), Ascent Flight Director John Shannon said, "Yikes!"

The main engines controller replied, "You bet!"

The booster officer said, "Concur."

John concluded, "We don't need any more of these!"

This unlikely combination of failures actually worked to our advantage. On our right engine, we had the fuel leak, and one controller wasn't functioning because of the momentary power loss. Had both right engine controllers been working, they would have pumped more fuel into the engine to try to compensate for the leak. That would have resulted in our running out of hydrogen before we ran out of oxygen. If the fuel ran dry, the engine could have exploded.

We leaked more than twenty-five hundred pounds of hydrogen through the ruptured cooling tubes. In a lucky break, we were accidentally short-loaded more than eight hundred pounds of oxygen due to a tanking error before the launch. The engine cutoff sensors saw that the oxygen tank was running dry and shut down the engines early. Thank goodness those sensors did their job. (We'd be cursing those cutoff sensors in my next mission.)

The center engine, also operating without one of its two computer controllers, had a slight bias in one of its pressure transducers. That caused the engine to start running "fuel lean," extending the

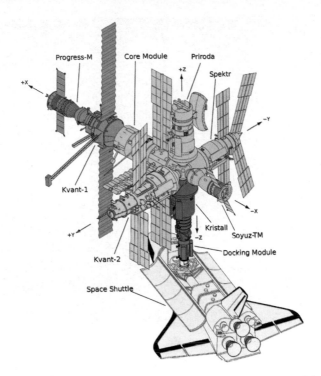

The Shuttle–Mir complex as it was from late 1996 through the end of the Shuttle–Mir program. (Diagram by Orionist, Creative Commons via Wikimedia, based on a NASA diagram, accessed at https://commons.wikimedia.org/wiki/File:Mir-Shuttle_diagram.svg)

Mike Foale, Vasily Tsibliyev, and I enjoy a group meal with the shuttle and Mir crews in the Mir base block module on the first day of our docked operations. (NASA)

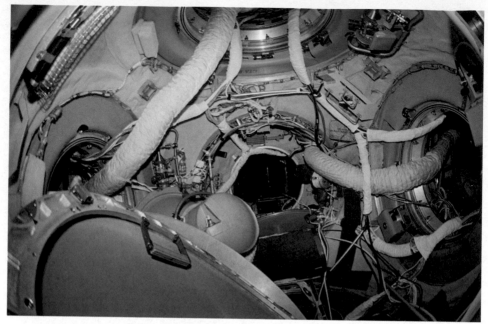

The interior of Mir's central node, showing the bewildering labyrinth of hatches, ductwork, and cables among the station's modules. Commander Tsibliyev placed the red fabric around the hatch at right to help the STS-84 crew find our way back to *Atlantis* in case of emergency. (NASA)

Our group photo in SPACEHAB during our last day of docked operations. In gray shirts from lower left: Jerry Linenger, me, Ed Lu, Jean-François Clervoy, Elena Kondakova, Carlos Noriega, and Charlie Precourt. From left in Mir flight suits: Vasili Tsibliyev, Aleksandr Lazutkin, and Mike Foale. (NASA)

Looking eastward over the Mediterranean Sea toward Greece.
The "boot" of Italy is at left. Views like these always inspired
me to think of the thousands of years that brave explorers and
merchants sailed through these waters. (NASA)

With President Bill Clinton in the Roosevelt Room of the White House on
March 5, 1998, for the announcement that I would be the first woman to command
an American space mission. (NASA)

The STS-93 crew patch, showing the Chandra X-Ray
Observatory exploring the mysteries of the universe. The
French flag was for our crewmember Michel Tognini.
(NASA)

An oddly coincidental and unusual location for an STS-93 crew
photo—with one of the space shuttle main engines. Cady Coleman
and I are in the engine bell; Michel Tognini, Jeff Ashby, and Steve
Hawley are in front. (NASA)

The Chandra telescope and IUS (white-and-gold cylinder at bottom) were installed in *Columbia*'s payload bay while the space shuttle was "in the vertical" at the launch pad. The curved silver panels on either side are *Columbia*'s payload bay doors. The two small windows at the top are the aft flight deck windows, through which we observed checkout and deployment of the telescope on Flight Day 1. (NASA)

Donning my harness and joking with the closeout crew in the White Room
before entering *Columbia*. (NASA)

Sitting in the commander's station (and lying on my parachute) during Terminal Countdown
Demonstration Test (TCDT). On launch day, I was strapped into my seat like this during the
last several hours of the countdown. (NASA)

Postlanding photos of
Columbia's damaged right
engine. A gold pin was ejected
from the engine at ignition and
punctured three tubes on the
inner chamber wall that carry
hydrogen to cool the engine
bell. Had the pin punctured two
more of the tubes, the engine
would likely have exploded
and killed us. (NASA)

The drogue chute and speed brakes slow us down on the runway following our night landing at Kennedy Space Center, July 27, 1999. (NASA)

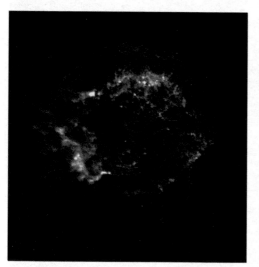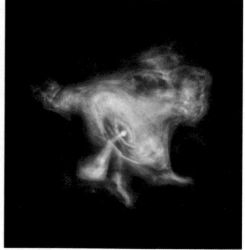

At left is Chandra's first image, of Cassiopeia A, a shell of stellar wreckage formed by a violent supernova explosion about three hundred years ago. The dot at the center of the image is the neutron star left after the star exploded, revealed for the first time. At right is the whirlpool of high-energy particles produced by the rapidly spinning pulsar at the heart of the Crab Nebula. Low-energy X-rays are in red, medium-energy X-rays are in green, and high-energy X-rays are blue. (NASA/CXC/SAO photos)

NASA's Space Flight Awareness program reminds all of NASA's employees and contractors that astronauts are real human beings with families who depend on everyone doing their job as perfectly as possible. This photo of me and Bridget was taken the day after STS-93 landed. (NASA)

We changed the STS-114 crew patch after the *Columbia* accident to incorporate the space shuttle-shaped patch from *Columbia*'s final mission, as our homage to Rick Husband and his STS-107 crew. (NASA)

Charlie Camarda, Andy Thomas, and I review procedures in the Virtual Reality Lab at Johnson Space Center before beginning a training session. (NASA)

Physical training is important no matter what your role is on the crew. Here I'm lowering myself from a shuttle hatch during emergency egress training at Johnson Space Center's mockup facility. (NASA)

The STS-114 crew in the White Room prior to our mission. Back row, left to right: Charlie Camarda, Wendy Lawrence, Soichi Noguchi, and Andy Thomas. Sitting or kneeling: "Vegas" Kelly, me, and Steve Robinson. (NASA)

The "kid pic" drawn by the children of *Discovery*'s crew in the hours before launch. It still hangs at Kennedy's Launch Control Center, outside room 4P23. (NASA)

STS-114 was my only daylight launch! We lifted off on July 26, 2005, at 10:39 EDT and were in space eight and one-half minutes later. This was the first shuttle launch after the thirty-month stand-down following the *Columbia* accident. (NASA)

The aft of the flight deck is a very crowded and busy place during "proximity operations," as we rendezvous and dock with the ISS. I'm in the pink sweater, flying the approach. The ISS is out our overhead windows. (NASA)

Sergei Krikalev on the ISS photographed *Discovery*'s belly as we executed the first-ever rendezvous pitch maneuver. This allowed the ISS crew to check our ship for damage. The view at right shows the logistics module in our payload bay. (NASA)

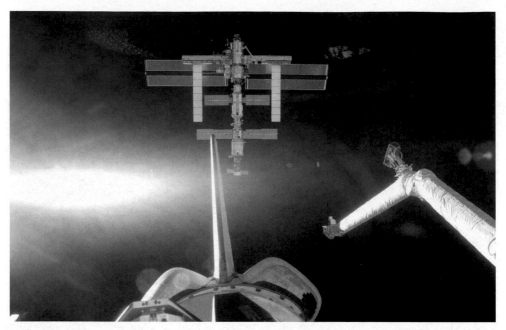

The International Space Station gleams in the light of the rising sun about six hundred feet behind *Discovery*'s tail as we execute our slow turn in the pitch maneuver. (NASA)

I shot this photo from the flight deck's far right window as Steve Robinson rode on the Canadarm 2 during the third spacewalk of the mission. (NASA)

Taking a short break in the Russian segment of the ISS. (NASA)

The STS-114 crew joins Launch and Entry Flight Director LeRoy Cain and the flight controllers and support personnel from Mission Control to celebrate the successful conclusion of our mission. Thousands of dedicated NASA and contractor employees across the country helped to ensure our mission's success. (NASA)

Being reunited with my family was always the best part of returning from a flight. This photo was taken in the KSC Operations and Checkout Building in February 1995 after my return from my first flight, STS-63. My father (Jim Collins) is at left, and my mother (Rose Marie Collins) is at right. My brother Jim is in the white shirt next to my mother, and my sister, Margy, is in the middle, with her husband, Ed Conklin, standing behind her. Family friend Fr. Bill Zamborsky is in the upper right.

Celebrating a birthday with Pat and the children in 2001. I retired from NASA in 2006 to spend more time with my family.

burn time on that engine. Had the short circuit not occurred and the center engine been burning fuel at its normal rate, we would have been as much as two hundred feet per second short of our required velocity at engine shutdown. That wouldn't have been fast enough to get into orbit, leading to a high-speed abort at night into whatever trans-Atlantic abort site we could reach—perhaps not a survivable situation. Again, we were lucky.

I feel slightly embarrassed at how happy—even clueless—my voice sounded on the radio when I said, "It's great to be back in zero-g again," given what was going on in Mission Control. On *Columbia*, we honestly had no clue about how close to disaster we'd come.*

Sending Chandra on Its Way

We opened the payload bay doors and immediately began preparing Chandra for deployment. Cady and Michel powered up Chandra and the IUS, while controllers in Cambridge and Sunnyvale remotely activated and checked out the systems on the spacecraft.

Five hours after we were in orbit, Cady rotated the tilt platform up to twenty-nine degrees and paused for another status check. Everyone said the IUS and Chandra looked healthy, so Cady issued the commands to disconnect the umbilical cables. She then rotated the tilt table to point Chandra at a fifty-eight-degree angle from *Columbia*. At 7:47 a.m. (EDT), Cady pulled the DEPLOY switch. Springs pushed Chandra out of the tilt table at less than one-quarter mile per hour. We watched through the overhead windows in the cockpit as Chandra soared silently above us. I gently backed us away with a few pulses of *Columbia*'s thrusters.

---

* Flight controllers use the recording of that whole eight-and-a-half-minute ordeal as a case study for training. You can watch an annotated recording of the Mission Control audio loops at youtube.com/watch?v=PdN1rWaXxh8. Wayne Hale also wrote an excellent technical analysis of all the things that went wrong—and right—in his blog at waynehale.wordpress.com/2014/10/26/sts-93-we-dont-need-any-more-of-those/.

One hour later, Sunnyvale commanded the first stage of the IUS to fire. Chandra was on its way to its new orbit.

The primary goal of our mission was a complete success, and we hadn't even finished our first day aloft. I read an email from Mission Control with news about the fuel leak, but that was behind us now. I went to sleep thinking that the rest of the mission would be a piece of cake.

Flight Day Two: Chaos

The crew set up and began operating the secondary experiments. Two experiments needed to be videotaped, and one of our two video cameras failed. We could not repair it. Neither crew member wanted to give up running their experiment, so we developed a plan to share the camera.

The experiment I most looked forward to ended in frustration. Astronauts love SAREX, the Space Amateur Radio Experiment program. (We call it ARISS these days, for Amateur Radio on the ISS.) We use ham radio technology to communicate with school students on Earth. It's highly motivational for the astronauts, as we get to take a break from our other tasks and talk with children from across the country who are interested in space. And, of course, the students enjoy speaking with astronauts in space! The sessions have to be tightly scheduled, as we are only within communications range of a participating school for about ten minutes.

When we reached the appropriate time in our schedule for the first session, I called down to the school. (My call sign was KD5EDS.) Despite repeated calls, my voice wasn't going through. Ten minutes came and went without my being able to connect, and then we were out of range. I knew the students on the ground were disappointed that they couldn't talk to us. I went back over the checklist. I discovered that I'd missed a step and left a circuit breaker open.

Our crew was overscheduled as it was, and we all fell even further behind as the day progressed. We kept bumping into one another in the mid-deck, where many of the experiments resided. Meanwhile, we all had to perform basic maintenance tasks to keep *Columbia* healthy. And Houston kept calling up changes to the flight plan.

I hit my redline for tolerance of chaos and my maximum limit of frustration. I knew the crew wasn't happy, either. At the end of the day, I called a crew meeting in the mid-deck.

"This is not the way we are supposed to operate," I said.

I told the crew that we needed to be talking more to one another and less to Houston. If there was a problem, we needed to put together a plan on board *Columbia* first, and then call the recommendation to the ground. I asked each of the crew their suggestions to add more organization for the next two days.

CAPCOM Story Musgrave called up another change from Houston in the middle of the meeting.

"Don't answer," I told the crew.

Story called again a few minutes later, and again I told the crew to ignore the call until we finished our discussion. We were radio silent for thirty minutes.

After the mission, I explained the situation to Story. "I knew exactly what was happening," he said. "I understand." As a veteran of six space shuttle missions, he had firsthand experience with frustrations resulting from a lack of resources, insufficient time, and too many changes.

Leaders need to call a time-out whenever things begin spiraling out of control. Whether you're in a basketball game or in the midst of a crunch to meet a deadline, a leader must have the courage to call a halt and help the team regroup. It's important to ask everyone what they're seeing and what help they require to be successful.

I certainly learned valuable lessons for my next mission. My primary role as commander was to ensure that we could execute the flight plan successfully. If not, take items out of the plan or provide more resources and time. Rehearse running all of the secondary experiments in an integrated way, not just separately. Don't wait until you're in orbit to discover that the schedule puts people in the same place at the same time. Have a process for reprioritizing activities if necessary.

These seem like simple principles, but it's easy to forget them in the heat of the moment. Taking time for realistic planning up front prevents errors and frustration later on.

The Second Half

Just after we entered orbit on launch day, and while I was still in my seat, Houston asked me to check cockpit control panel L4 to see if there were any open circuit breakers. Mission Control was trying to diagnose what happened to our AC power during ascent. Still wearing my helmet and pressure suit, it was difficult for me to turn my head to look over my left shoulder at the breaker panel. It was still dark outside when I scanned the breakers. I told Houston that all of them were closed.

When I floated up to the flight deck during our third day in orbit, I glanced over at the L4 panel from a different angle. To my dismay, I saw that the AC1 Phase A circuit breaker for the right engine was open. The collar of the breaker's sleeve is supposed to be white, but this breaker had been pulled so many times that the white paint was worn off. Now it was the same shade of gray as the metal panel. I hadn't seen the collar on the first day.

I wasn't sure how I should report it to Houston. I needed to think about what I was going to say.

That night, before turning in, I called down to Mission Control. "I have something I need to talk to you about, and I need the EGIL and the main engine folks to listen in."

They asked me to stand by as they gathered the backroom teams specializing in electrical and instrumentation and the shuttle's main engines.

I explained that the circuit breaker was in fact open, and that I had missed it due to lighting, the worn collar, my position, and so forth.

They weren't upset. In fact, they said, "That's great! Thanks a lot. What you just told us confirms what we already figured out. Every other indication was that we had a short in AC1 Phase A."

I'm still surprised that I missed that circuit breaker. Fortunately, it didn't affect the safety of *Columbia* or our crew. I tried to put it out of my mind and concentrate on our other work.

My favorite secondary experiment was SWUIS (pronounced "Swiss"), the Southwest Research Institute Ultraviolet Imaging System. The SWUIS principal investigator was Dr. Alan Stern, who subsequently entered the history books as the principal investigator for the New Horizons mission to Pluto and beyond. SWUIS was a seven-inch Questar telescope specially modified to detect ultraviolet light. Earth's atmosphere absorbs UV light, so we can't study the sky in that wavelength from ground-based telescopes.

Steve Hawley attached the telescope to the window on the shuttle's hatch—the only window on *Columbia* that didn't filter out UV light. I turned *Columbia* to point the window toward Steve's targets, then maintained a stable orientation by "collapsing the deadbands" of our autopilot to minimize the amount of allowable drift so that the objects stayed in the telescope's field of view. Steve took video of our Moon, Venus, and Jupiter and its moons. He searched for evidence of vulcanoids, which are hypothetical asteroids inside the orbit of Mercury. To our great disappointment, we didn't find them.

As a professional astronomer, Steve was certainly in his element. He spent hours trying to conduct as many observations as he could during our short mission. I began to worry that he was overworking himself. I told him several times, "Steve, relax and enjoy the view!"

I was able to reschedule the SAREX radio call to the school I missed earlier in the mission. That was a relief. "Mom" wanted everyone to be happy, especially the kids.

Cady ran an experiment with genetically engineered plants. They produced certain chemical signatures if they were stressed in space, by lack of oxygen or disorientation from lack of gravity. This research provided valuable insights for future long-duration missions, when astronauts will have to depend on food they can grow in space.

Our other experiments involved everything from caterpillars (which later became butterflies for study by schoolchildren) to single-celled organisms and even prototypes of hinges for the solar arrays to power the International Space Station.

I wish we'd had more time in orbit. I had lobbied unsuccessfully for a longer mission over a year earlier but eventually relented for the good of the program. On the fifth day of our flight, it was time to pack up and come home.

Everyone completed their fluid loading protocol an hour before our de-orbit burn. Once again, I couldn't get down all twenty-four ounces of the salty liquids I was supposed to drink. I ensured everyone was strapped into their seats before the burn. Once you begin the reentry process, events proceed rapidly. You must be seated and ready.

Our track took us over Texas at ten o'clock at night on July 27. We treated the residents of the state to a spectacular light show as we zoomed 200,000 feet above them at Mach 12, trailing a long streak of pink-purple plasma from our encounter with the upper atmosphere. *Columbia* went into a right bank to shed energy, and Jeff could look straight down at the city lights of Texas and Mexico.

"There's Mexico City!" he exclaimed.

I can imagine what a glorious view he had from his windows. Mine were facing up into the sky as we banked.

Merely ten minutes later, we crossed the Florida coast. I took over control from the autopilot as *Columbia* became a subsonic glider. Despite thousands of practice approaches in the Shuttle Training Aircraft and the simulator, this was the first time I was actually flying a *real* space shuttle through the atmosphere. I found it easier to fly than the Shuttle Training Aircraft. *Columbia* responded to my control inputs smoothly and without delay, with superb handling qualities. We sailed south along the Space Coast of Florida with a view of

Cocoa Beach straight ahead, made a turn to the right, and lined up for final approach to the northwest on Runway 33.

This was only the twelfth night landing of the shuttle program. Of course, I had trained extensively for landing in the dark. However, it's always an interesting experience when you can't make out anything on the ground that isn't brightly illuminated. The computer image projected on our heads-up display lined up nicely with the floodlit runway.

I was just one knot off our targeted touchdown at 195 knots indicated air speed. We had a slight gust of crosswind from the right, but not enough to throw us off. Jeff deployed the drogue chute to slow us down. Soon, we called, "Wheels stop."

The tour bus drivers at Kennedy Space Center tell everyone nowadays that I made the most precise landing at KSC in space shuttle history. I don't know if that's true, but who am I to argue with them?

Having stayed in space less than one week, we experienced no problems readapting to Earth's gravity. We powered down *Columbia* to turn her over to the ground crew and then bounded down the stairs to inspect the ship. We all saw for the first time the damage to the cooling tubes in the right engine.

I spent the night in a hotel with my family and flew back to Houston the next morning. Before we left KSC, a photographer snapped a picture of me carrying three-year-old Bridget. NASA later used that photo in a safety poster for the Space Flight Awareness program. Space Flight Awareness reminds NASA and contractor employees of the importance of emphasizing quality and safety in everything we do, since *real people's* lives are on the line.

Vice President Al Gore headlined our welcome-home ceremony at Ellington Field. After we expressed our thanks to the crowd and left the platform, I saw Lisa Reed. I ran over to her and showed her my crew notebook.

"Lisa, you're not going to believe this! Look at the last run you guys scripted. You sent us the exact failures we had on the way uphill! It was just like our last sim!"

Yet another coincidence.

Meanwhile, *Columbia* was in her hangar at Kennedy to be de-serviced and prepared for her trip to Palmdale. Technicians discovered a two-inch section of scorched bare wire under a panel in the payload bay. That wire had vibrated against a stripped screw head during twenty-five launches and landings over the past eighteen years. The insulation finally wore through, creating a bare wire that electrically arced against the screw when our mission launched. That momentary short circuit knocked two engine controllers offline.

Further inspections found more wires that were in danger of shorting out. NASA inspected the other orbiters, identifying thirty-eight sections of hazardous wiring on *Endeavour* and twenty-six on *Discovery*. NASA grounded the entire fleet for repairs. No space shuttles flew again until December 1999.

Technicians from KSC eventually presented me with the section of burned wire as a souvenir.

Postflight

The weeks immediately after a mission follow a predictable pattern: Medical exams. Writing the crew report. Reviewing and commenting on all of the photographs. Putting together the crew movie. Seemingly endless face-to-face technical debriefs on every aspect of the mission. Postflight press conference. I was up writing reports until well after midnight every night, grabbing sleep when I could, and then back awake before six o'clock.

My happiest memory of that period was actually the day after I returned to Houston. Pat needed to start flying for Delta again, so I took over the parenting role.

I put Bridget in her stroller and walked her around the golf course. How great that was! I had just come back from a monumental achievement—being the first woman to command a space shuttle mission—and now I was in the traditional role of a mother, taking my toddler for a leisurely stroll. I was indescribably happy. I reflected on the mission. It wasn't easy. It required a lot of work, and the circumstances weren't necessarily ideal, but I truly felt capable of balancing this amazing career with raising a family.

I gave forty-seven press interviews the first week I was back. Yahoo! News ran a story that some people at NASA were referring to me as "Janeway," after Captain Katherine Janeway of the starship *Voyager* in the *Star Trek* TV series. (Another strong woman of Irish descent!) They intended it as a compliment for the "cool manner" in which I handled the ascent to orbit. As far as I was concerned, our performance was due to the training team's excellent work in preparing our crew.

NASA released Chandra's "first light" image on August 26. It revealed the neutron star at the heart of the supernova remnant Cassiopeia A. Sometime about AD 1680, a massive star collapsed and exploded. It left behind an expanding shell of gas and dust, and a dense, city-size star made of neutrons. This stellar corpse is only visible in X-rays. Chandra's dramatic image of it was perfect.

Despite Chandra's design lifetime of five years, it is still generating valuable science more than twenty-two years after deployment. Chandra has shown us how stars are born and how they die, peered into the nuclei of galaxies to find supermassive black holes, and revealed the dynamics and structures of galaxy groups. I feel immensely proud to have contributed to the advancement of scientific knowledge by placing this amazing instrument in orbit.

NASA sent our crew on the road for public relations appearances. These started with a visit to NASA Headquarters in Washington. We next conducted a briefing on Capitol Hill, after which we met with individual senators and representatives to thank them for the work done in their states and districts to support the space program. We visited all of the major NASA Centers and the Smithsonian Astrophysical Observatory office in Cambridge that managed Chandra.

I participated in a wide range of media events celebrating our mission. I have lost count of the total number, but I made eleven appearances over the next several months in New York City alone.

The NASA Public Affairs Office scheduled these events to raise awareness of the space program with people representing all interest groups and backgrounds. We shared our experiences with our customers—the American people. It wasn't about us in particular, other than that we wanted folks to know that astronauts are just ordinary people who get to do extraordinary things. If you think it's about your own glory as an astronaut, then you're in the wrong job.

Cady and I flew to Los Angeles to appear as guests on *The Tonight Show with Jay Leno* on August 16. Pat and Bridget were with us backstage. Jay had a blast playing with Bridget, chasing her around the room and up and down the hallway. He told us which questions he wanted to pose to us on air and asked if we were comfortable talking about them. Jay was a pleasure to talk with during the program and was obviously knowledgeable about the space program.

I rang the opening bell at the New York Stock Exchange on January 7, 2000. I was on Oprah Winfrey's show on April 25, 2000. I appeared on the *Today* show, *Live with Regis and Kathie Lee*, and other morning shows. The tour felt like a whirlwind.

Early in the fall of 1999, Cady told me that she was pregnant. Of course, I felt elated for her. I found out shortly afterward that I was expecting, too, with a due date in June. I kept quiet about it at work, only telling my boss, secretary, and scheduler.

I miscarried in November, less than three months into my pregnancy. My doctor told me that it was nothing that I did wrong and that it wasn't because of my health. Rather, my age just made me statistically more likely to miscarry.

Next, I talked to my NASA flight surgeon, Richard Jennings, who was understanding and supportive. He told me, "I recommend you not fly for a week."

I objected. "Why? I feel fine."

He said, "I think you're okay physically. You just need to take some mental time off."

I didn't listen to him. That night, I flew to Kennedy in a T-38 and then flew the Shuttle Training Aircraft. I didn't need to take time off to sulk at home and feel sorry for myself. Flying took my mind off everything. Controlling the aircraft, solving problems, being in the moment: this was my safe space.

To end on a happier note, it wasn't long before I was expecting again. My son, Luke, was born in November 2000. I had a wonderful husband, two adorable children, and three space missions under my belt. Could life get any better?

Chapter 13

THE *COLUMBIA* TRAGEDY

Astronauts spend the majority of their careers performing desk jobs. Of the 5,768 days I worked for NASA, I spent a grand total of thirty-eight days, eight hours, and twenty minutes in space. That represented only 0.66 percent of my total NASA career.

Following STS-84, I became chief of the Astronaut Office Vehicle Systems Branch. Eighteen astronauts and engineers reported to me. As in any new role, I began with a fresh look at the scope and expectations. Were there opportunities to effect changes, even subtle ones, to improve how we worked to achieve our mission?

One of my first acts in this assignment was to request that we change the name to "Spacecraft Systems." The term *vehicle* didn't accurately reflect our full scope, which now included the International Space Station as well as the space shuttle. I ensured that there was an astronaut tracking every module under construction for the ISS as well as the avionics upgrades for the space shuttle. I held that role until NASA assigned me to STS-93, at which point I began mission training.

I served as chief information officer (CIO) for the Astronaut Office after STS-93. Perhaps the most controversial issue of that role was implementing Charlie Precourt's mandate that we convert from our labor-intensive, manual scheduling system to a fully automated calendar. Even a seemingly minor information systems change can actually involve a major cultural change. Resistance to change is part

of human nature. We had to reassure our excellent schedulers that
they would still be important and valued members of the team after
we implemented the new system. The process reinforced my belief
that leaders must practice clear vision, consistent communication,
patience, and deep listening if change is to succeed.

Next, I served as the shuttle branch chief. I represented the Astronaut Office at all of the program-level meetings chaired by Ron
Dittemore, the space shuttle program manager.

While Ron's meetings were important, they often lasted forever.
On Thursday, November 2, 2000, I was in the seventh hour of one
such meeting at Johnson. I was nine months pregnant. After several
uncomfortable hours of sitting, I hit my limit. I turned to my deputy,
astronaut Duane "Digger" Carey, and said, "I'm leaving. Cover for
me!"

My son, Luke, was born two days later.

I returned from maternity leave in January 2001 to a new role
as chief of the Safety Branch. I replaced Rick Husband, who left to
train with his crew for STS-107. Rick couldn't wait to move with
his crew into the office next door.

The Safety job had a huge scope. I represented the Astronaut
Office on safety issues regarding the shuttle, ISS, payloads, and our
T-38 fleet. As when I had assumed my other roles, I intended to
begin my tenure with a review of the effectiveness of our processes
and procedures and develop a strategy. I was almost immediately over-
taken by events, however. I seemed to spend all my time receiving
and following up on complaints and simply wasn't able to get ahead
of them all. I felt it was important to listen carefully to each person's
perspective on the issues, whether serious or silly. One person said I
needed to tell Bill Shepherd—the commander of the International
Space Station—to stop turning somersaults in zero gravity!

My job required me to travel to Kennedy for every mission's
Flight Readiness Review and the L-2 briefing two days before a
scheduled launch to ensure that any safety concerns had been properly addressed. Fortunately, I was eventually able to hire the superb

engineer Kevin Mellett from Kennedy as my deputy. I wouldn't have made it without his help.

My Final Mission

NASA announced that I would command STS-114, scheduled for late 2002. It would be a relatively routine mission—fly *Atlantis* to the ISS for crew rotation and to transfer equipment and supplies. Joining me on my crew were pilot Jim "Vegas" Kelly, Japanese astronaut Soichi Noguchi as MS 1, and Steve Robinson as MS 2. Flying up with us to take residence on the station were the ISS Expedition 7 crew of Ed Lu, Yuri Malenchenko, and Aleksandr Kaleri. We would bring home Ken Bowersox, Don Pettit, and Nikolai Budarin following their extended stay on board ISS.

I told a few close friends that I expected this to be my final mission. I had accomplished everything I came to NASA to do, except to visit the International Space Station. After I flew STS-114, it would be time for me to move on and give other pilots the opportunity to fly and command. Pat and I planned to sell our house and move to Florida. He could continue to fly for Delta from Orlando, and perhaps I could fly a desk at Kennedy Space Center.

I turned over my Safety Branch role to astronaut Dom Gorie in late 2001 so I could start training with my crew.

I almost immediately became concerned about the sheer amount of work scheduled for our flight. We had to transfer all the equipment and supplies from the 1,100-cubic-foot multipurpose logistics module to the station. That's equivalent to moving the contents of a three-bedroom house. Then we'd fill that module back up with other experiments, unneeded equipment, and trash to bring back to Earth. Steve and Soichi would take three spacewalks to install an external stowage platform on the station's Quest airlock and replace a gyroscope unit on the ISS that had catastrophically failed.

With the experience of STS-93 still fresh in my mind, I clearly saw we had too much work, insufficient resources, and too little time. I knew firsthand what happens when you fall behind trying to execute an impossibly tight schedule: you rush to make up for lost time—and that's when you make mistakes.

Facing tremendous schedule pressure at the highest levels, NASA was under a time crunch to finish assembling the US portion of the ISS. To keep costs under control, NASA's administrator mandated completion of the US portion of the space station up to the second node module (now named Harmony) by February 2004. That was going to require more than one dozen shuttle flights, with almost no room for schedule slippage. Even in the spring of 2002, many of us felt it would take a miracle to meet the 2004 deadline.

Technicians discovered cracks in the flow liners in the propellant feed lines of *Atlantis* in June 2002, then found similar cracks on *Columbia* and *Discovery*. NASA grounded the entire shuttle fleet for repairs until October, throwing the ISS assembly work even further behind.

NASA pushed my STS-114 launch date to March 6, 2003—less than five weeks after Rick Husband and his STS-107 crew were due back from their science mission aboard *Columbia*.

I felt bad for Rick and his crew. As a SPACEHAB mission plopped in the middle of the rush to assemble the station, STS-107 was almost a distraction for NASA. Except for a Hubble servicing mission, STS-107 was the last non-ISS flight on NASA's launch manifest until well after completion of the ISS. Rick's launch date had been pushed back more than ten times. Now it felt like the Program Office just wanted to get STS-107 over with.

Losing My Friends and *Columbia*

In December 2002, as my launch date approached, I felt the need to talk to seven-year-old Bridget about what her mom was about to do. She only vaguely remembered my previous mission, when she was just three. I was scheduled to fly not long after the annual remembrance of the *Challenger* accident of January 28, 1986. I wondered if one of Bridget's classmates might tell her about *Challenger* and scare her with the thought that her mom was going to blow up on the way to space.

I asked her if she had ever heard of *Challenger*. She hadn't.

I sat down with her and opened a book to a photo of the *Challenger* crew and the explosion. I said, "I want you to hear this from me, not from a kid at school." I explained about the accident, told her about each of the *Challenger* astronauts, and promised her that the cause had been completely fixed.

"Don't worry about me when I fly to space. I will be safe," I assured her.

Only six weeks after that conversation, we lost *Columbia* and her crew.

In the early morning of Saturday, February 1, 2003, I was at home with Luke, now two years old. Pat and Bridget were on a camping trip with the Indian Princesses father-daughter program at Huntsville State Park, northeast of Houston.

I woke up Luke. "The space shuttle is coming home!" I said excitedly. "Let's watch the space shuttle landing on TV." Of course, he had limited understanding. Since he would soon be watching me go up on my own mission, I was trying to get him to comprehend as much as possible in his own terms about the space shuttle and what his mom would be doing.

As we sat on the couch and watched the NASA-TV feed from Mission Control, I held a space shuttle model. I demonstrated to him, "This is what Mom will be doing next month!" We listened to the STS-107 crew execute the de-orbit burn, and everything seemed to be going well. Their two-week mission had been so flawless that most people had forgotten a space shuttle was in orbit.

I was excited to see STS-107 come home, because once they landed, my crew would become the Prime Crew. As the next mission to fly, we would have priority throughout the shuttle program.

Columbia flew its hypersonic entry starting at Mach 25. The potential energy stored in the orbiter during its ascent converted to heat as the vehicle began to descend. To slow from 17,500 mph to landing speed, *Columbia* slammed into the air molecules in the tenuous upper atmosphere and executed banking turns to dissipate its pent-up energy. Portions of the shuttle's external surfaces reached almost 3,000°F (about 1,650°C) during reentry.

I listened intently to the radio calls, knowing the shuttle's maneuvers through this phase of its flight. All seemed fine until *Columbia* began crossing Texas. The radio seemed to stop working, and I was surprised the crew wasn't answering the standard calls.

"*Columbia,* Houston, comm check," CAPCOM Charlie Hobaugh called out. "*Columbia,* Houston, comm check." Switching to another radio system, he called, "*Columbia,* Houston, UHF comm check. *Columbia,* Houston, UHF comm check."

No answer.

My sense of concern went into overdrive when the tracking station controller at Kennedy Space Center said he did not have a radar return on *Columbia.*

My heart started beating so hard that I thought it would jump out of my chest.

Something was terribly wrong. The shuttle is always exactly where it should be, as it has no capability to speed up or slow down. Its trajectory speeds are fixed. If it wasn't where it was supposed to be, then . . . *where?*

I started changing channels. I stopped on a major news station, and what I saw horrified me: burning pieces of debris streaking across the Texas sky. Reports of explosions were coming in from around the areas crossed by the shuttle's ground track.

Oh no. My friends could never survive this.

Seven astronauts were lost. Our shuttle *Columbia*—lost.

As my heart pounded harder and harder, I grabbed a stool and took Luke to the kitchen sink. I never let him play with the kitchen sink faucets. This time was different. I needed to keep him occupied while I made urgent phone calls.

First, I called my mother. I told her I was okay and not to worry about me. "And by the way," I said, "I will not be launching on

March 6." Next, I called my father, who wasn't aware yet there had been an accident. He seemed bewildered as to how something like this could happen, but he was surprisingly calm.

I can only imagine what my parents were feeling about the families of *Columbia*'s crew.

Next, I started calling my crew members, beginning with my pilot, "Vegas" Kelly. He had already heard. He sounded focused. He told me he was leaving one of his children's sporting events to head into work. He was the family casualty assistance calls officer (CACO) for one of the *Columbia* astronauts, Laurel Clark. Every crew member designates a primary and secondary CACO before a mission. Before your flight, you ask yourself the sobering question, "If something bad happens to me, which of my colleagues do I want to look after my family?" The CACO's solemn responsibility is to advise and assist the next of kin in the event of a horrible loss like today's. My mission specialist Steve Robinson was the secondary CACO for *Columbia* astronaut Mike Anderson.

I called my crew one after another, primarily to ask if they had heard about the accident. I told them, "I don't know what the cause was. You can call me back anytime."

I also called the members of the space station crew who were to fly up to the ISS on my mission. Later that day, Ed Lu informed me that he already heard he would be moving to Russia in the coming week to prepare for a launch on a Russian Soyuz spacecraft. Ed and his Russian colleagues would continue their mission as the next space station crew. But now, the only launch vehicle available to them was the Soyuz.

I couldn't reach my husband. He had no cell phone service. Pat later phoned to say that word of the accident was spreading fast, and pieces of debris might be on the ground in the heavily forested area near their campground. He canceled the second night of his stay and headed home with Bridget.

When Bridget returned, she refused to talk to me about the accident. I tried to explain to her about the risks of spaceflight, that we control those risks, and that this was truly an accident.

"Sweetie, I don't want you to worry about your mom. I am not going to space unless all these problems are fixed."

She wouldn't listen. The accident hit her hard, and she deeply internalized it. I was devastated. I'd lost the trust of my seven-year-old daughter. It took years to rebuild our relationship.

Aftermath

Pat stayed home to watch Luke, and I drove to Building 4 at Johnson to meet with the other astronauts. Kent "Rommel" Rominger called a meeting of our office at five o'clock, just after he returned from Kennedy. Rommel had been out at the runway that morning, waiting for *Columbia* to land. He and Bob Cabana had the awful duty of informing the families of *Columbia*'s crew that their loved ones were lost.

I caught up with Rommel in the parking lot outside our building just before five o'clock. We walked in silence.

As we approached the door, he finally said, "You wouldn't believe the day I had today."

What more was there to say?

Rommel's leadership impressed me. The essence of his message at our meeting was, "Don't point fingers. Don't make assumptions. We don't yet know what happened. We all have our ideas. It's important we work together with the shuttle program. We'll solve this as a team."

NASA's administrator Sean O'Keefe had appointed an investigation board late that morning. Admiral Harold W. Gehman, the former NATO supreme allied commander, was its chair. His board included some of the most respected people in aviation and space, including Sally Ride, who had also served on the *Challenger* investigation board in 1986.

Having flown military aircraft in relatively risky jobs for the past twenty-five years, I had seen my share of fatal accidents. You never get used to them. You grieve the loss of life, the tough job of helping the pilot's family through the ordeal, and the internal questioning of how this possibly could have happened.

It also hits you: *This could have happened to me.* There is a distinct aura or feeling of the unknown—a fear, even a creepiness—that settles over you until every aspect of the tragedy is uncovered and analyzed.

Although Rommel told us not to make assumptions, I had a theory. Right after I saw the blazing streaks across the Texas sky, my first thought was that one of *Columbia's* auxiliary power units (APUs) had blown up. In my first NASA job, I tracked APU problems for the astronauts. There were several failure modes, one being the highly volatile and explosive propellants that power the APUs. During reentry, the APUs burn like engines and power the hydraulic systems that drive the flight controls. The three APUs sit relatively close to one another. If one exploded, you could lose all three. The ship would tumble out of control. This was in my head all day Saturday.

Accident investigations start with a *fault tree* diagram. You write down every possible cause as "branches" of the tree. You eliminate possible causes as the investigation progresses and evidence is gathered. The APUs were certainly on the fault tree. By the end of the day Saturday, we were also hearing speculations about foam.

Foam? I wondered. *Wait a minute, I think I read something about foam in a shuttle program report a few weeks ago.* I searched my email and was unable to find any relevant hazard reports.

Several of my 1990 astronaut classmates and I met for dinner that night at Mely's Mexican restaurant in League City. We sat with our spouses and talked about how we would support the *Columbia* families and the office. We wondered about what our roles would be over the next weeks, months, and years.

Had I still been the safety officer, I would have deployed immediately to the accident scene. I'm not sure how we could have managed that, with Pat's flight schedule and my family situation. My replacement, Dom Gorie, relocated to the command center at Lufkin, Texas, on the night of the accident and led the debris recovery effort until the end of April. Jim Wetherbee also went to Lufkin to coordinate the search for *Columbia's* crew. Mark Kelly and several other astronauts deployed to the field to assist local search teams over the next two weeks.

Soichi and I were the only two of the STS-114 crew remaining in Houston who didn't have another critical assignment. We spent most of that chaotic first week after the accident going to meetings and talking to people. I attended many funerals and memorial services for my friends.

Rommel assigned a group of four astronauts, including me, to be office spokespeople and answer questions from the media. Reporters pushed me to the brink of frustration at times. I had to remind myself that they were just doing their jobs. However, I did become visibly upset one time when a journalist kept hammering away on a sensitive topic during an interview. She later apologized to me.

———

Much to my shock and disbelief, NASA Headquarters asked me to continue to train for my upcoming mission. They initially told me to work toward a March 6 launch but shortly thereafter moved it to April 7. The official word was "Train for a launch one month from now, three months from now, or six months from now."

I knew there was no way this could possibly happen. *Never.* The country would never allow another space shuttle to fly until we identified and corrected the cause of *Columbia*'s demise. Besides, two of my crew members would be unavailable for weeks while they helped the families of *Columbia*'s crew.

Headquarters insisted we keep training, though. The April launch date moved to May, then July, then September. The unrealistic planning frustrated our crew. Even worse, it wasted effort and resources. Operational personnel throughout the system had to develop schedules and work toward every planned launch date. If you're going to fly in May, a *lot* has to happen in March and April to make that happen, as there are critical events that must occur at specified times before a launch.

Vegas kept asking me, "Why don't they just acknowledge reality and move the launch date out a year?" It would have prevented significant organizational turmoil.

———

In such a chaotic situation, what was I supposed to do now? I felt helpless—one of the worst feelings of all.

We sincerely wanted to assist in the field. I also didn't want to distract people or impede the recovery effort. This was not about shining a spotlight on ourselves. Being on-site for even a single day would help my crew heal our own feelings, honor *Columbia*'s crew, and support the thousands of people who were searching for debris.

Our crew flew to Nacogdoches, Texas, on April 10. We toured the debris collection hangar and saw some of *Columbia*'s wreckage recovered so far that week. We joined a twenty-person crew on a grid search, methodically walking slowly across the landscape for several hours while looking for anything unusual. Soichi found a piece of tile from *Columbia*'s underside. That was a profound moment for all of us.

We traveled in May for a dinner at the Lufkin Civic Center celebrating the completion of the debris-recovery effort. We thanked the residents of East Texas for their incredible support and talked about NASA's plans for the International Space Station.

Hard Lessons for a Flawed Culture

The days and weeks went by, and the branches of the fault tree were pruned back as the investigation eliminated possible failure scenarios. We learned that the physical cause of the accident appeared to be the presence of a hole, about six to eight inches wide, on the leading edge of *Columbia*'s left wing. The leading edge is made of reinforced carbon-carbon (RCC), a hard but brittle material intended to protect the spacecraft from the high temperatures of reentry.

How did that hole get there? *Foam?*

On January 16, eighty-one seconds after liftoff, a piece of foam fell off the orange external tank and hit *Columbia*'s wing. The sprayed-on foam insulates the tank from the humid Florida air. Without it, the super-cold fuel and oxidizer inside would cause the air's moisture to freeze into a layer of ice on the outside of the tank. The ice could break off during the ascent and damage the shuttle's heat shield.

In this case, the technology designed to protect the shuttle actually harmed it. The foam piece that broke off the left side of the fuel tank was about the size of a briefcase and weighed less than two pounds. How could that lightweight piece of foam cause so much damage to a hard heat shield? Basic physics says that even if something is lightweight, if it is going fast enough, its momentum (the product of mass times velocity) packs enough energy to cause significant damage. *Columbia*'s wing impacted that piece of foam at a relative speed of more than 500 mph.

Columbia entered orbit with a hole in its wing that could not possibly be repaired—one that was absolutely certain to doom the shuttle.

And no one knew about it.

The Columbia Accident Investigation Board published its report in August 2003. It was the most well-written and thorough accident report I've ever read. It did not limit its findings and recommendations to the physical cause of the accident. The most surprising and disappointing conclusion was that the organization and culture at NASA were as much to blame for the accident as was the foam.

NASA faced a huge set of challenges: fix the problems that caused the accident, get the shuttle flying again as soon as safely possible, finish building the ISS, and then replace the shuttle with the next-generation human launch vehicle. The Board said we wouldn't be able to accomplish any of those if we didn't first fix the flaws in NASA's culture.

What? How can that be? I work in the best culture anywhere!

I was comfortable working at NASA. As a woman, and as the first woman pilot and commander of a space shuttle, I always felt fully accepted. I was able to do my job without being distracted from the mission at hand.

Here's the catch: when you're in a leadership position, you don't always see the cultural problems. You have to seek them out.

I seriously pondered the NASA culture and my role in it—how I might inadvertently be reinforcing bad practices, and how I might use my influence to help effect positive change. I've summarized here what I learned over years after the accident as we tried to help NASA once again be the best it could be.

Respected leaders—astronauts in particular—need to be aware that people in the organization will always try to please you, especially when they are working on your mission. They naturally want to tell you how well things are going. In an organization that by its nature depends on taking risks, you need to hear more than just the good news. To learn what's actually going on, you have to listen hard, seek out people's concerns, and avoid intimidating anyone. I realized that my "feeling comfortable" may not always be the desired state.

I thought about the classical principles of leadership I learned in Air Force officer training: (1) know your job well and do it with excellence; (2) know your people and communicate with the team; and (3) integrity first.

Knowledge of your job, or your technical ability, means executing your role with competence and excellence. As a shuttle commander, my knowledge could save my life and the lives of my crew. I studied every day, even if it meant reading worksheets while I pushed my children in their strollers. Our crew would practice various situations and set up challenging competitions in our simulator runs. I would memorize procedures, talk with engineers about the nuances of their systems, and insist our instructors keep throwing tough problems at us in training. As a highly technical institution, I believe that NASA, its contractors, and our teams excelled in knowledge of our jobs. (We will see later that this can be a strength overused.)

The second classic leadership principle is to know your people. This includes your colleagues, your "customers," and everyone else who supports your mission. For each of your team members, you should know their name, their role, their talents, interests, family situation, and special needs.

Our astronaut crews were well trained in teamwork. We knew one another so well that we could usually predict one another's actions. Most of our tasks were too complex for one person to do them alone.

Integrity means honesty. It is more than simply telling the truth. Integrity is trusting someone will be honest with you in the future. Rather than keeping secrets from one another, you must share critical knowledge. A leader with integrity sets standards for herself and maintains them for everyone equally. When such a leader makes a decision, the team knows the leader's motivation: it is to do what is best for the mission, not for personal gain.

Our shuttle crews, training teams, and Mission Control embodied these principles, but what was missing from the larger organization? Breakdowns in how NASA handled the *Columbia* accident were painfully obvious when the whole story was revealed.

Engineers watched the postflight video of *Columbia*'s ascent the day of the launch. They saw the chunk of foam fall off the fuel tank and hit *Columbia*'s left wing. The view was not optimal; they only saw a puff of shattered debris spray off the impact site. They couldn't directly determine if there was any actual damage to the spacecraft.

Concerned, the engineers requested that the intelligence community use their reconnaissance assets to photograph the shuttle in space and look for damage. The Air Force said they would gladly work on the request.

Later that day, due to a variety of communication breakdowns, organizational mix-ups, and misunderstandings, NASA management withdrew the request, even after the Air Force had agreed to help.

Had NASA known there was a breach in the wing, we could have mounted an all-out effort to launch a rescue mission. I believe we likely could have saved the astronauts, even if *Columbia* herself was doomed. The rescue mission would have captivated the world's attention. People would have been riveted watching NASA bring the *Columbia* astronauts safely home. It would have surpassed Apollo 13.

"NASA to the rescue, thrilling the world." NASA at its best.

Instead, the investigation board highlighted numerous missed opportunities resulting from faults in NASA's organizational culture. The board defined *organizational culture* as the "basic norms, beliefs, and practices of an institution. At the most basic level, organizational culture defines the assumptions employees make as they carry out their work; it defines 'the way we do things here.'"

Culture is the lens through which we see problems, and it is the spoken and unspoken rules about how we work and interact with one another. We need to be aware of culture's influence and shouldn't let the culture alone drive our decisions.

The dysfunctions pervading our culture led to flawed thinking that prevented us from responding appropriately to the situation. I want to emphasize that I'm not pointing fingers at anyone. I have been guilty of this kind of thinking. It takes insight to recognize where culture is impacting our thinking, and it takes courage to challenge the status quo. Unfortunately, it sometimes takes a rude shock like an accident or an unexpected failure to force people to step back and examine errors in their thought processes.

What does flawed thinking sound like?

First, a senior engineer told me, "A lightweight piece of foam cannot put a hole in a hard heat shield. It is physically impossible." This is where knowing you're the expert gets in the way of openness to other possibilities. That engineer was proven wrong in July. A test team at the Southwest Research Institute loaded a one-pound piece of foam into an air cannon and shot it at 500 mph at a mock-up of a space shuttle wing. The foam blew a hole in the wing about the size of the one suspected of dooming *Columbia*. The investigators called the test the "smoking gun." People who witnessed the test still had a hard time believing what they had seen. Was that a lack of creative thinking? Even experts can be wrong, and they still have much to learn.

Next, we heard the excuse that our software models didn't predict any damage from foam strikes. Analysts used the only tool at their disposal, a software program called CRATER, to try to determine if the foam strike would cause serious damage. However, NASA never validated CRATER by real-world testing on strikes as big as the one suffered by *Columbia*. CRATER was a good model for small hits, but you couldn't necessarily extrapolate the results to a major impact.

Garbage in, garbage out.

How many times did we hear in meetings over the years, "It's never happened before, so it won't happen now"? The space shuttle was a delicate system that had managed to survive many impacts and close calls throughout the previous twenty-two years. Debris impacts during the STS-27R ascent to orbit caused more than seven hundred damaged tiles on *Atlantis*, but the vehicle still made it home. That led to overconfidence. "We've always survived this kind of thing before."

The problem was that the design rules required that there not be *any* impacts on a space shuttle during launch. *None.* The fact that we'd survived impacts before was totally irrelevant. We had discounted and failed to correct the *real* problem.

Because we'd gotten away with it before, we accepted the risk and ignored the rules. The more we cheated disaster, the less likely we thought disaster was. Sociologist Diane Vaughan coined the phrase "normalization of deviance" in her 1996 book on the *Challenger* launch decision. When a system is operating outside its design parameters, that should be a red flag, not an excuse to relax the rules.

After the accident, I was shocked to learn that on STS-112 in October 2002—two missions before STS-107 and three before my scheduled flight—a piece of foam fell off an external tank. It peeled off from the exact same location as the hunk of foam that doomed *Columbia* and smashed a six-inch dent into a metal electronics control box on a solid rocket booster. NASA briefly mentioned the issue in the next mission's flight readiness review and said it was being investigated. The STS-112 incident didn't even come up in the flight readiness review for STS-107.

We weren't dodging bullets. We were playing Russian roulette.

And then there was the pale justification, "Even if we knew there was a hole in *Columbia*'s wing, there was nothing we could do about it." That was akin to the old proverb, "If you can't stand the answer, don't ask the question." True, there was no plan available at the time—but how could anyone say that we couldn't have *developed* one?

Creative thinking by some of the smartest engineers in the world would have come up with something. At least we should have let them try. Instead, we completely closed the door and trusted in blind luck, rather than giving *Columbia*'s crew a fighting chance.

It was heartbreaking to think about all the ways in which we let down Rick Husband, Willie McCool, Mike Anderson, Laurel Clark, Kalpana Chawla, Dave Brown, and Ilan Ramon, not to mention their families.

We could never bring them back again. But we could take a hard look at ourselves and thoughtfully consider how to change NASA's culture to reduce the chances of our making the same kinds of mistakes. We identified several basic leadership practices that people could employ at any level of the organization, in every meeting and interaction.

First and foremost was to improve our listening skills. We needed to train ourselves to listen to others with an open mind, not letting our additional motivations (like schedule or budget pressure) color how we heard what people were telling us.

Humility was an important partner to listening skills. This was not in the sense of being quiet and shy. Rather, a humble but strong leader realizes, "Maybe my way is not the only way. What if I gave someone else's ideas a chance?"

Creativity was another mental muscle we needed to build through exercise. Some of our return-to-flight challenges were so daunting that it appeared we might not be able to meet them if we merely improved existing hardware or practices. We needed to innovate, to question underlying assumptions. That was sometimes tough in a "mature" program like the space shuttle, which had been in operation for twenty years.

One example was the requirement to inspect the space shuttle's heat shield in orbit. Someone came up with the brilliant idea of having the space shuttle flip, nose over tail, in a full circle so that the space station crew could photograph the shuttle's belly and check for damage. I was in the meeting where that idea first came up. Some engineers argued that this could not possibly work. They quoted a flight rule: once the shuttle was within one thousand feet of the ISS, the shuttle crew had to keep the ISS in constant sight. That wouldn't be possible if the shuttle flipped over.

After I'd taken a moment to reflect, I realized, *We wrote the flight rule; we can change it*. We couldn't let our normal way of operating keep us from considering creative options. We had to be willing to challenge the status quo if necessary.

Listening, humility, and creativity contributed to our having disciplined teams—ones with a sense of responsibility for the mission and respect for the people we worked with.

The Space Shuttle Program used the concepts of "probabilistic risk management" for systematic and quantitative assessment of the risks inherent in our extremely complex engineering system. We had to remind ourselves that while this approach was a great tool, we always needed to ensure that our culture wasn't leading us down the wrong path.

No matter how good our culture and our analyses were, we would never be absolutely perfect in our execution. After all, we were human beings, and no matter how well-intentioned we were, we could never be flawless. We needed to work continuously on our culture and our individual practices.

———

And so, NASA began the arduous process of assessing the necessary hardware and process changes to make the shuttle safer to fly.

Notice that I did not say, "*Safe* to fly."

The investigation board's report highlighted something that we all knew: the space shuttle would never be 100 percent safe. Since the orbiter attached to the *side* of its fuel tank and boosters, the heat shield was exposed to potential damage, and there could never be a practical ejection capability for the crew. There would always be situations where a major systems failure would lead to the loss of the vehicle and its crew. That's why all the new space vehicles have returned to the concept of Mercury and Apollo, with the crew sitting in a capsule on top of the booster, and with an escape system that can pull them quickly away if anything goes wrong.

We were operating on the knife-edge between life and death, between marvelous engineering and the harsh laws of physics. We always needed to stay on guard.

Symbolizing Hope

In the wake of the accident and the investigation board's recommendations, NASA changed key leaders in the Space Shuttle Program Office. These significant and influential roles had oversight over all aspects of the program, including manufacturing, maintenance, testing, and operations. As the return-to-flight commander, I coordinated closely with all these key leaders.

I realized that I had a much broader leadership role as a *symbol* of why NASA was undertaking the monumental task of fixing the flaws in its culture and returning the space shuttle to flight. My crew and I represented the hopes and aspirations of the entire agency and all of its contractors. NASA needed the space shuttle to fly again, and we were the human faces the world would see on that spaceship. Our husbands, wives, children, and parents would be trusting everyone in NASA and its contractors to fix the space shuttle's flaws and bring the STS-114 crew safely home again.

People would be watching us very closely over the coming months, looking for any signs of hesitancy, doubt, distrust, or fear. They would see how we conducted ourselves in meetings. We needed to embody the NASA culture of the future, to lead by example.

I postponed any thoughts of retirement and relocation to Florida. I couldn't possibly carry through with that now in good conscience. People might interpret my departure as a lack of faith in NASA and its people.

I committed to stay on and command the return-to-flight mission, no matter how long it took us to get it off the ground.

Chapter 14

RETURN TO FLIGHT: STS-114

In the fall of 2003, NASA announced that we would not fly STS-114 until the shuttle program addressed the *Columbia* investigation board's recommendations. We didn't know how long it would take for the shuttle to be ready to return to flight. The fleet had stood down for 975 days between the 1986 *Challenger* accident and the 1988 STS-26R return-to-flight mission. I anticipated we'd also need more than two years to recover from *Columbia*.

Return-to-flight (RTF) missions after an accident are highly significant in the aerospace industry. We had to implement critical improvements to technologies and processes, and the only way to test them end to end was with a real crew flying a real mission. We needed to prove as astronauts, an agency, and a nation that we were ready and willing to resume flying the space shuttle.

If our mission failed, it would undoubtedly mean the immediate termination of the Space Shuttle Program.

What was previously a routine crew rotation and logistics mission now became a demanding and complex developmental test flight. Our primary objectives included resupplying the space station, testing new flight safety techniques and procedures, and evaluating technologies for making minor repairs to a shuttle's heat shield in orbit.

We retained our four "core" team members: me as commander, Jim "Vegas" Kelly as pilot, Soichi Noguchi as MS 1, and Steve Robinson as MS 2. The three seats previously designated for crew rotation would now be filled by astronauts dedicated to our mission's key objectives.

Kent Rominger and I identified the new crew members, whom we announced on November 7. Australian American astronaut Andy Thomas would be our intravehicular (IV) astronaut, coordinating the mission's complex spacewalks. That was originally one of my responsibilities. My role as RTF commander had grown far too complex for me to take on the IV duties, as well.

Wendy Lawrence would "fly" the space station's robot arm. She'd manage the immense task of transferring five thousand pounds of materials and equipment from the shuttle and logistics module to the ISS, and six thousand pounds from the ISS for us to carry back to Earth. Andy and Wendy were two of NASA's most experienced mission specialists.

I also wanted to fly another rookie on the mission. We selected Charlie Camarda, a brilliant engineer who worked as a research scientist at NASA's Langley Research Center before becoming an astronaut in 1996.

Training manager Dale Williamson, lead instructor Juan Garriga, and I developed our training plan. For the first time, I had a cell phone during crew training, and it was a godsend! Being able to call Dale and Juan from anywhere maximized the efficiency of our training time. I monitored our training and attended operational meetings with Paul Hill, the lead flight director, and LeRoy Cain, the ascent and entry flight director.

Wendy and Vegas trained on the space station's robot arm. Andy and Charlie practiced with the shuttle's arm. Steve and Soichi spent considerable time training for their spacewalks at the Neutral Buoyancy Lab's huge underwater facility. Vegas and I rehearsed proximity operations, rendezvous and docking procedures, flying the shuttle training aircraft, and sims of our launch and entry flight profiles.

Wendy developed the detailed protocols for transferring the materials between the logistics module and the ISS, with Charlie's assistance. The last time a shuttle visited the ISS was December 2002.

Even with a reduced ISS crew of only two, basic supplies and consumables were running low. We also needed to exchange equipment and materials supporting the ISS's scientific experiments between the shuttle mid-deck and the station.

Hardware Fixes

Wayne Hale, the deputy manager of the shuttle program, met with our crew regularly. I can't say enough about how fantastic he was. He always asked, "I'm here to listen. What's on your mind? What are you concerned about?" When we realized we wouldn't fly for at least a year, he asked me, "Now that we have time to make hardware changes, what would you like to see fixed?"

I had several concerns. First was an issue with studs in the eight posts that held the space shuttle to the launcher platform prior to liftoff. At the instant the solid rocket boosters ignite, electrical signals fire explosives to rupture the nuts securing the studs to the launcher. When the studs are ejected down into the hold-down posts, the shuttle is released to fly free. All eight nuts have to break at precisely the same time, and the studs have to fall away cleanly. If one or two studs don't release properly, a shuttle can still lift off. But if three or more hang up, the shuttle might keel over and explode.

Twenty-three previous launches experienced "stud hang-ups." The second shuttle launch in 1981 and STS-92 in October 2000 had multiple stud hang-ups. This dangerous performance issue cropped up on 20 percent of all previous flights but was never addressed. Talk about normalization of deviance!

NASA began redesigning the detonator system after our request, but it wasn't ready in time for our mission. Engineers assured us that there was an acceptably low probability of this issue recurring on our launch.

My second concern was the booster separation motors (BSMs) on the solid rocket boosters. BSMs were small solid-fuel engines to push the expended boosters away from the space shuttle two and one-half minutes into flight. Investigations after the *Columbia* accident showed that BSMs could be a source of debris. NASA redesigned them for our mission.

Engineers added sensors behind the leading edges of the wings. These could detect impacts like the one that doomed *Columbia*.

Finally, I was concerned about the four rudder/speed brake (RSB) actuators in the space shuttle's vertical tail. These mechanisms drove the hydraulics that turned our rudder, which also fanned open to act as a speed brake to manage our energy during our final approach and landing. I had read inspection reports about corrosion on the shuttle's body flap and elevon actuators, undoubtedly from the salty Florida air. No one had inspected the RSB actuators, though. They were hard to reach, requiring dismantling and rebuilding the heat shield on the tail. Management pushed back on my request, saying that it could take a year to perform the inspection on *Atlantis*. However, engineers already planned to inspect the RSBs on *Discovery*, which was currently down for maintenance.

Their inspection revealed a shocking and potentially fatal problem. In addition to corrosion and cracking in *Discovery*'s RSB actuators, a gear in the uppermost actuator was installed *backward,* as was one in a replacement actuator. In every mission since 1984, that actuator was trying to move *Discovery*'s rudder and speed brakes in the opposite direction than was being commanded. Had a commander forcefully "stomped on the rudder" to compensate for unexpected crosswinds during a landing, the rudder could have jammed, causing the vehicle to go out of control.

NASA immediately ordered the replacement of the RSBs on *Atlantis*. Parts from *Endeavour*'s RSBs were later pulled, inspected, and installed on *Atlantis* to supplement the incomplete spare parts on hand.

Discovery would be ready to fly sooner than the other two shuttles, so *Discovery* swapped places with *Atlantis* as our orbiter for STS-114. *Atlantis* would be the "Launch on Need" rescue vehicle if anything went wrong during our mission.

—

My crew frequently visited the NASA Centers and our contractors' facilities to support their work efforts and hear their questions, concerns, and ideas. At Marshall Space Flight Center in Huntsville and at the Michoud Assembly Facility outside New Orleans, we talked to engineers and technicians who were trying to fix the foam loss problem on the shuttle's external tank.

At Michoud, I examined the tank assigned to my flight. Engineers redesigned the structure where the foam fell off *Columbia*'s tank. I wondered about redesigning or eliminating other debris sources, including the protuberance air loads (PAL) ramps, foam-covered fixtures nearly forty feet long that prevented airflow from getting underneath cable trays and flow lines during ascent. An engineer explained, "If you catch any air under there, it could break and cause an explosion. We scraped off the old foam and reapplied new foam."

I had similar concerns about the foam on the "ice frost ramps." I was told that those manually applied areas of foam were vital to keep ice from forming on the bare metal brackets on the tank. Chunks of ice coming loose and hitting the shuttle during ascent would cause far more damage than foam.

I stopped pushing the issues. NASA discussed and approved the existing design at the program level. A report issued three months before our flight said the PAL ramps were safe to fly. However, the report noted that there was still a possibility of debris from the area. A design team was developing alternatives for future missions.

Bellows at joints along the liquid oxygen feed line were another potential source of foam loss. Engineers assessed that, although the bellows were safe to fly as is, beginning three missions after mine, the bellows units would have heaters.

New Techniques
Columbia reentered the atmosphere with fatal damage because her crew had no way to examine the shuttle's heat shield in orbit. Flight

Director Paul Hill led a team to develop ways for shuttle crews to inspect an orbiter's tiles, leading edges, chin panel, and nose cap to ensure the vehicle was safe to fly home from space.

Their first innovation was an extension boom for the shuttle's robot arm. We'd carry this aloft on the starboard sill of the payload bay. The forty-foot boom had three specialized cameras to closely inspect every square inch of the nose cap and the wing leading edges. We'd be able to see anything as small as a hairline crack. Vegas, Andy, and Charlie would survey the shuttle during Flight Day 2 to look for launch damage. They'd inspect the shuttle again for orbital debris damage before we came home.

The second new procedure was the Rendezvous Pitch Maneuver (RPM). I would fly an approach from behind and below the ISS as usual, aiming for a point six hundred feet from the ISS when we were along the *R-bar* line between the station and the Earth. At that point, I would stop the shuttle's motion in all axes. I'd then initiate a pitch-up command and engage the autopilot to fly a nose-over-tail flip at three-quarters of a degree per second. The crew on the ISS would photograph our heat shield with 400mm and 800mm telephoto lenses as we slowly turned. After the maneuver, I would fly four hundred feet in front of the ISS and then slowly inch in with a *V-bar* approach for docking.

Developing and flying the maneuver required considerable training and simulator time. Once we worked it out, the procedure was simple, elegant, and operationally executable. Shuttle commanders flew it on every mission (except for the Hubble servicing mission) from 2005 until the end of the program in 2011.

If the boom sensors or the RPM identified any significant damage, we had no assured way to repair our ship in orbit. We would dock with the ISS and empty *Discovery* of all her consumables. Once we were all back on ISS, we would undock *Discovery* and send her on autopilot to burn up on reentry over the South Pacific. The seven members of our crew would stay on the ISS as a safe haven for up to forty-five days until *Atlantis* could pick us up. If *Atlantis* was unable to fly for some reason, then the Russians would bring us back in three Soyuz flights.

I appreciated the concerted efforts made to improve the space shuttle system's hardware and safety procedures. Although not all issues were corrected, I firmly believed we were flying the safest space shuttle in the history of the program.

Preparing to Fly

Discovery rolled out to the launchpad on April 6, 2005. Technicians loaded *Discovery*'s external tank with liquid oxygen and liquid hydrogen for a tanking test on April 14. Two sensors in the liquid hydrogen tank failed. Those sensors would cut off the shuttle's engines if the tank ran short of fuel. Launch rules forbade our flying without the sensors working properly.

Our launch date was pushed from May 13 to July 13 to study the situation. NASA Administrator Mike Griffin insisted that heaters be installed on the tank's LOX line bellows during the delay.

Technicians swapped out the sensors and ran another test in May. The sensors passed, but the test uncovered other problems with the tank that technicians couldn't correct at the pad. *Discovery* rolled back into the Vehicle Assembly Building on May 26 to swap its external tank and solid rocket boosters for the ones intended for *Atlantis*. *Discovery* arrived back at the launchpad on June 15.

———

Against this backdrop of preparing my ship and crew to fly, I also had to prepare my family for my mission. Both of my kids were now old enough to understand on some level what I was about to do. They needed time for the situation to sink into their minds. Additionally, as the mission commander, I understood that the crew's family members would also need special preparation for this emotional flight. With the assistance of my training manager and lead instructor, we planned family flights in the shuttle simulator, visits to the neutral buoyancy pool, tours of the T-38 aircraft and shuttle training mock-ups, and group parties. I assured my children and the other crew families that I wouldn't fly this mission unless I was totally satisfied that any outstanding concerns about shuttle safety were addressed.

Our family watched a TV show about *Challenger* and *Columbia* a few months before my mission. After the show, Bridget asked Pat, "Why did the astronauts die?"

Pat explained to her, "Sometimes things happen. Every day, there are car accidents where people unfortunately pass away. Sometimes people walk across the road and something happens. You just never know."

Interestingly, Luke later asked Pat the same question, and Bridget repeated Pat's answer to Luke. Bridget seemed to understand, but I'm sure she was far from feeling comfortable about my rocketing off into space again.

I told her on another occasion, "You are going to be a little bit afraid when I fly this flight. That is just the way it is. That is human nature. I want you to know that is normal. I also want you to know that I would not fly this flight if I did not believe that I was going to be safe and come back home again."

My mother had a much harder time dealing with the stress. *Challenger* was already nine years past when I flew my first mission. She was nervous then but able to cope. *Columbia* was far too recent for her to put it out of her mind. Since her health was declining, too, she decided to stay home in Elmira for this launch.

—

Finally, it was time to enter quarantine and say goodbye, which for me was the hardest part about flying a mission. With any luck, this would be my final family farewell as an astronaut.

We all felt sad and a little tense. As we drove to JSC, I told myself the white lie that helped me weather these separations: *I'm not leaving my kids for three weeks. NASA might scrub the launch in the morning, and I can come right back home again tomorrow night.*

Pat pulled the car up in front of the quarantine facility. After I walked away, Bridget began sobbing uncontrollably, "Mom's leaving!" Bridget continued crying as they drove off. Four-year-old Luke finally had enough. He said, "Bridget, stop crying! You're gonna give yourself diarrhea!"

I wish I'd been there to hear that to lighten my mood!

The minute I walked into the quarantine facility door, I entered into my role as a shuttle commander—a role I was very comfortable with. And then all was well with me.

Our crew boarded the Astrovan at noon on July 13 for the seven-mile drive out to the launchpad. It was raining, and thunder occasionally rumbled during our ride. The forecasters assured us that the skies would clear well before our 3:50 p.m. launch time.

Funny—I never thought we would actually launch on July 13. It was an "unlucky" day, it was the first launch in two and one-half years, and there are always thunderstorms in Florida on July afternoons.

I strapped into the commander's seat on *Discovery* at 12:38 p.m. Steve was the last one to crawl through the hatch and take his seat at 1:15. The skies had cleared. We settled in and completed our first communications checks at 1:27.

Three minutes later, an engine cutoff sensor in the external tank failed. It was exactly the same problem that we thought NASA corrected in April. The team in the Firing Room immediately scrubbed the launch.

Unfortunately, the many residents and community leaders from East Texas who came to see the launch had to head home. Without their dedication in the search for *Columbia*'s crew and debris after the 2003 accident, we might never have been able to return the space shuttle to flight.

No one knew how long it would take to correct the fuel tank sensor issue. Our window of possible launch dates extended to July

31. One requirement for returning to flight was that we launch and separate from the external tank in daylight so we could photograph the tank to look for foam loss. Orbital mechanics precluded our launching in daylight between August 1 and September 8.

Assuming engineers immediately fixed the problem, our next launch date was three days from now. But we might wait as long as eighteen days, throughout which we would remain in quarantine.

On July 15, the launch date slipped until "at least late next week." I issued a statement on behalf of the crew that we would remain in quarantine and maintain our proficiency for the mission. I expressed our confidence in the ground team and our appreciation that NASA leadership was taking the time to understand and correct the problem.

I tried to keep our crew occupied to relieve the monotony. We toured historic launch sites at Cape Canaveral and drove around the Merritt Island Wildlife Refuge. We flew back to Houston to practice in the simulators. I eventually requested permission for us to break quarantine and visit our families, who were waiting for launch in the KSC area. I spent an afternoon at the ocean with my family and ate dinner at the Cocoa Beach IHOP.

That same day, engineers believed they isolated the issue with the fuel level sensors, but they weren't 100 percent certain. They set a launch date of Tuesday, July 26. We went back into quarantine.

Then weather became a concern. Tropical Storm Franklin formed near the Bahamas on Thursday, July 22. Its future ground track was uncertain. If it trended toward KSC, *Discovery* would have to roll back into the Vehicle Assembly Building. Fortunately, the storm eventually took a wide turn away from us.

At the L minus two-day management briefing, NASA's leaders took the unusual step of agreeing to waive the mission rule requiring four functioning fuel cutoff sensors —provided not more than one failed. Four cutoff sensors permitted redundancy. If one failed, the other three would still provide a safety margin.

We listened to dissenting voices and considered their objections. Everyone admitted to extensive soul-searching that they were not succumbing to launch fever, but rather making a prudent decision based on solid engineering data. NASA's administrator agreed with the recommendation to proceed with the launch countdown.

As with the foam situation on the external tank's PAL ramp, engineering analysis determined that we were at an "acceptable" level of risk.

Should I have demanded further study or testing? I had listened thoughtfully and soberly to the issues and resolutions. I heard enough to assure me that it was okay to launch. I wouldn't have put my crew's lives or *Discovery* on the line if I felt it was unsafe.

Return to Flight

We woke up at the crew quarters at just after midnight on July 26. When we gathered for breakfast and crew photos at five o'clock, we heard the uplifting forecast of an 80 percent chance of favorable weather for the launch.

We left the O&C Building for the launchpad at 6:49 a.m. When Soichi arrived in the White Room, he held up signs for the camera, GET OUT OF QUARANTINE FREE! and OUT TO LAUNCH.

The closeout crew shut and latched our hatch at nine o'clock. Vegas and I spent the next ninety minutes activating the myriad systems on the spacecraft, checking our computers, configuring displays, and running through a seemingly endless series of checklist steps.

Those ninety minutes race by for the commander and pilot. The wait must seem interminable for the crew on the mid-deck, who have nothing to do except count the minutes.

On the fourth floor of the Launch Control Center, the kids of *Discovery's* crew engaged in a long-standing tradition: creation of the mission's "kid pic." NASA provided colored markers to the children to decorate a large whiteboard with drawings and messages of encouragement and love. It helped the kids pass the several hours between their arrival at the LCC and our launch. The kid pics from all the missions still line the LCC's long hallways.

We entered the last countdown hold at T minus nine minutes, as the launch team in the Firing Room ran through their final checks. At ten-thirty, we heard the call from Launch Director Mike Leinbach we'd been anticipating for more than two years.

"Okay, Eileen, our long wait may be over! On behalf of the millions of people who believe so deeply in what we do: good luck, Godspeed, and have a little fun up there!"

I replied, "Our thanks to you, to the launch team, and to everybody in the shuttle program. The crew is go for launch."

The countdown resumed. Escorts guided our families to the roof of the LCC for a private and unobstructed view of our launch.

Discovery's three main engines ignited. The solid rocket motors fired six and one-half seconds later. We were on our way at 10:39 a.m.—coincidentally, the same instant as *Columbia*'s final launch.

An unlucky turkey vulture, circling over the pad at liftoff, smacked into and slid down the external tank before we cleared the launch tower. We hurt the vulture worse than it hurt us.

Engineers and the viewing public watched our ascent in unprecedented detail and from a variety of angles. Cameras mounted on the external tank and the solid rocket boosters documented any foam loss. Two WB-57 reconnaissance aircraft circled at high altitude to film us as we passed by. A new C-band radar antenna at KSC and two shipborne X-band radars monitored the critical first few minutes of our flight.

The ride uphill felt exceptionally smooth and seemed uneventful, especially compared with my previous launch. After we entered orbit, we separated from the external tank. I pitched *Discovery* over so that we could take images and video of the tank as it tumbled away.

Engineers on the ground immediately analyzed the launch video from the external tank's video camera and were dismayed at what they saw. A tile chip came off one of our nose wheel doors. Most disturbing, several chunks of foam separated from the external tank. A few small fragments came off the ice frost ramps. A large piece, more than three feet long and one foot wide, broke off the PAL ramp and flew just under *Discovery*'s right wing shortly after the solid rocket boosters separated. It was about the size of the one that doomed *Columbia.* Sheer luck prevented it from impacting us.

NASA had clearly not fixed the foam loss issue.

We could continue our mission, but NASA immediately grounded the space shuttle fleet. No space shuttle could fly again until engineers fully understood and corrected the foam problem.

Mission Control told us about the situation six hours after we entered orbit. Our crew was understandably disheartened. However, we were safe for now. We had an unprecedented capability to inspect *Discovery* thoroughly and stay at the ISS if necessary. I reminded the crew that all we could do at the moment was focus on our mission, rather than worry about the unknown.

———

On Flight Day 2, Andy, Charlie, and Jim undocked the Orbiter Boom Sensor System and spent seven tedious hours scanning our nose cap and wing leading edges. The survey didn't reveal any significant damage, just a few small dings in tiles. A slightly billowed insulation blanket under the commander's window looked unusual. Engineers assured us it was not a risk for reentry.

The next day, we rendezvoused with the ISS. It was my turn in the spotlight, flying the first rendezvous pitcharound ever attempted. Sergei Krikalev and John Phillips photographed *Discovery*'s heat shield from the ISS. All went just as we had trained. The video of our pirouette was mesmerizing.

As commander, I manually flew *Discovery* four hundred feet in front of the station and lined us up for a *V-bar* approach. (All shuttle rendezvous maneuvers were always hand-flown by the commander, unlike modern spacecraft, which fly this maneuver on computer control.) With our nose pointed into space and our open cargo bay facing the ISS, we inched in slowly. At 7:17 a.m. (EDT) on July 28, *Discovery* docked. We were the first American ship to visit the ISS in 969 days.

Compared to the cramped, humid, and well-used Mir, the ISS was roomy, bright, new, and relatively uncluttered. Sergei and John conducted our safety orientation on the ISS. Almost immediately afterward, we went to work. Wendy and Vegas used the station's robotic arm to grab the shuttle's sensor boom on the right side of the payload bay, move it around and past the docking apparatus, and hand the boom over to the shuttle's arm for continued examination of *Discovery*'s heat shield.

———

During our first full day at the station, Wendy used the station's arm to remove the logistics module from *Discovery*'s payload bay and dock it to Node 1 on the station. Meanwhile, Vegas and Charlie used the shuttle's arm and sensor boom to inspect six areas of *Discovery*'s heat shield in greater detail.

Andy and I conducted interviews for three Sunday morning news shows. I expressed my surprise about the foam loss. Andy added that we were disappointed from an engineering standpoint, because so many people had worked on the issue but not yet licked the problem. We both agreed that the shuttle fleet had to remain grounded until the situation was resolved, to avoid putting future crews at risk.

I was already starting to think—and I even alluded to it in the interview—that we must start designing a spacecraft to replace the space shuttle after completion of the International Space Station. We needed the shuttle to complete the ISS, but it was simply too risky to keep flying the shuttle after that.

Steve and Soichi took the first of their three spacewalks the next day. In *Discovery*'s cargo bay, they tested several possible repair techniques on samples of damaged tiles and a reinforced carbon-carbon panel, applying the "goo" with devices that looked like caulk guns and spatulas. Then they installed the base mounting unit for an external equipment stowage platform on the space station's airlock module.

During the remainder of their six-hour, fifty-minute space-walk, they also replaced a GPS unit and rewired one of the station's control moment gyros (CMGs). The ISS has four CMGs, each of which contains a metal plate that spins at 6,600 RPM. These massive gyroscopes maintain the station's orientation using electrical power rather than rocket engines. Each CMG is about the size of a washing machine and weighs more than six hundred pounds. CMG-1 had failed catastrophically in June 2002, and CMG-2 was experiencing intermittent power problems.

Two days later, Steve and Soichi were back outside again, this time for more than seven hours, to remove CMG-1 and replace it with a new unit that we brought up with us. Soichi and Steve successfully replaced the gyro and reconnected it. They locked down the failed unit in *Discovery*'s payload bay to return to Earth for analysis.

The final spacewalk of the mission began differently from our original plans. While analyzing the photos taken during the rendez-vous pitcharound, engineers discovered two gap fillers protruding between tiles on *Discovery*'s belly. The gap fillers were about the size and thickness of an index card. Somehow, they had worked loose during launch. They stuck out far enough to create airflow disturbances that might cause an issue during reentry. They had to be removed. For the only time in space shuttle history, NASA was going to send an astronaut underneath a space shuttle during flight.

I can't say enough about the incredible work done by the back-room teams in Houston to prepare for this unusual spacewalk. They designed the complex arm movements to get Steve where he needed to go without bumping into our heat shield. They spent an hour reviewing the procedures with Steve, Soichi, Wendy, and Vegas. Wendy and Vegas then trained for forty-five minutes on the new arm movements using a simulator program on the ISS. That Wendy and Vegas could execute such a complex and unrehearsed process with so little preparation time demonstrated the wisdom of our approach to being experts in our roles: "Train for the *skill*, not for the *task*."

The unprecedented views of *Discovery*'s belly from Steve's helmet cam were simply stunning. Steve rode the end of the station's robot arm to the gap fillers, which he was able to pull out using only his gloved fingertips. Then Steve joined Soichi for the remainder of the planned tasks for the EVA outside the ISS. These included installing the external stowage platform and an experiment pallet to expose various materials to the space environment for an extended time.

The Mission Management Team extended *Discovery*'s stay at the ISS by an additional day. We transferred additional materials from the shuttle to the ISS. No one knew when the next space shuttle would fly again. Mission Control instructed Wendy that we should leave behind for the ISS crew anything extra that we had on the shuttle— laptops, cameras, nineteen bags of water, LiOH canisters for removing carbon dioxide from the air, food, tape, clothing.

After the years of intense work leading up to this mission, I enjoyed experiencing life on the ISS. I was able to "fly" around in the various modules and experience an incredible sense of freedom. I could run like a hamster in the cylindrical logistics module. When it was empty, I could run around the inner walls, keeping my head in one place, while my feet ran the circle. In the laboratory module, I was able to do every type of gymnastics maneuver you could dream of, by grabbing the bars on the ceiling, turning and spinning, extending my arms like a skater, and learning new ways to move in microgravity.

While running a checklist in the ISS airlock, I was floating around with dozens of bags surrounding me, so I couldn't see the ceiling, walls, and floor. Without my being aware of it, my body rotated and shifted while I worked. When I looked up from my task, I instantly found myself feeling disoriented and detached, similar to being in a state of vertigo. I recalled a similar experience on my first mission, so this time it felt more entertaining than stupefying. My perplexity went on for several minutes, until I was able to find the hatch and exit to the station. While I was enjoying these antics and surprises, I could not help but think of the *Columbia* crew; I am sure they all had similar wonderful experiences in their SPACEHAB module.

On the morning of August 4, the shuttle and ISS crews conducted a joint tribute to fallen astronauts and cosmonauts. Andy Thomas wrote a beautiful and touching ceremony that celebrated the cause of exploration and remembered those who gave their lives in extending humankind's reach into space. After reflections by each of us in turn, crew members repeated these words in English, Russian, and Japanese:

> *To the crew of* Columbia, *as well as the crews of* Challenger, *Apollo 1, and* Soyuz *1 and 11, and to those who have courageously given so much, we now offer our enduring thanks. From you we will carry the human spirit out into space, and we will continue the explorations you have begun. We will find those new harbors that lie out in the stars and of which you dreamed. We do this not just because we owe it to you, but we do it because we also share your dream of a better world. We share your dream of coming to understand ourselves and our place in this universe. And as we journey into space you will be in our thoughts and will be deeply missed.*

I closed with the prayer often spoken for those who sacrificed themselves for all of us:

> *They shall not grow old, as we that are left grow old:*
> *Age shall not weary them, nor the years condemn.*
> *At the going down of the sun and in the morning*
> *We will remember them.*

Coming Down for the Last Time

We were scheduled to return to KSC on Monday, August 8. We felt absolutely confident that we had the most thoroughly inspected heat shield possible. I knew without a doubt that we would have an uneventful reentry.

Since we hadn't had the opportunity to fly the Shuttle Training Aircraft for several weeks, Vegas and I practiced our approach and landing while we were still in orbit. A flight simulator program (PILOT) loaded on one of our laptop computers could read the inputs from the shuttle's actual controls and produce a relatively high-fidelity simulation of the reentry and landing. The sim provided us with the opportunity to reacquaint ourselves with the checklists and all the events that would occur in rapid succession once we rolled out on final approach. Jim and I tested *Discovery's* thrusters and control surfaces to ensure the ship was ready.

The weather forecast looked decent for Florida, with good visibility and only a slight chance of showers in the vicinity. We packed up everything on Sunday night. The next morning, we suited up, drank our fluid loading, and awaited the "go" for our de-orbit burn. Houston waved us off, as the clouds over Kennedy weren't cooperating. Houston waved us off again one orbit later, our last landing opportunity for the day.

I know the crew felt frustrated by the wave-offs and the hassle of fluid loading and getting in and out of the suits. I tried to demonstrate a positive attitude that I hoped would help the crew. I said, "Hey, we're just fine here. We've got another day in space, so we might as well enjoy it!"

We spent much of the wave-off day with "window time." We were so busy at the ISS that we scarcely had time for anything other than our assigned tasks. Steve played music on his laptop as we hung like bats from the mid-deck ceiling. We watched the Earth roll by. On one pass, when our ground track took us across the full breadth of Australia from northwest to southeast, Andy gave us a narrated tour of his home country.

Conditions didn't look any better for Kennedy the next day, so after two more frustrating wave-offs, we prepared to land at Edwards Air Force Base. Here was an opportunity to make a grand return to the home of my alma mater, the Air Force Test Pilot School!

Discovery flew its nighttime reentry flawlessly over the Indian and Pacific Oceans. The shuttle's maneuvering jets fired, their reflections in the plasma making it appear as if we were flying through a thunderstorm. The flight felt strangely smooth in contrast to the visual excitement, and we watched the colors melt from orange to yellow to green.

Soichi, Charlie, and Wendy sat in the mid-deck and couldn't see the exterior view, so, to keep them informed of our progress, I called out Mach numbers as we decelerated. I called out Mach 20, which we all knew was the point in *Columbia*'s reentry when the orbiter began to break up. We were in a completely different location—out over the Pacific rather than over Texas—but this was the speed at which *Columbia* lost control. To us, the "Mach 20" call was the equivalent of "Go at throttle up" for *Challenger*.

I knew with absolute confidence that we were going to be fine. I also knew the folks on Earth were holding their breath for us. They may not have been aware of all that we'd done to maximize our safety. They could only recall what happened to *Columbia*. I sensed absolutely no concerns among my crew about our own safety.

We came in toward Runway 22 at Edwards a few minutes after five in the morning Pacific time, while it was still dark. Two runway aids for instrument landings—the precision approach indicator (PAPI) lights and the ball bar—seemed unusually bright. The dry air and high altitude made them appear much brighter than the lights at Kennedy. Because I couldn't make out the ball bar light just prior to touchdown, I landed using the heads-up display inside the cockpit and touched down slightly earlier and faster than I intended. I'm glad I made a good landing on STS-93, because I wasn't happy with my touchdown speed on this mission.

Nobody else could tell that it wasn't the world's best landing. However, I was upset that I hadn't thought to ask Edwards to set a dimmer approach light setting. From a crew performance perspective, that was the only discrepancy during the entire flight.

For Vegas, Wendy, Charlie, Andy, and me, this was the final landing of our astronaut careers.

We powered down *Discovery* just as the sun rose. The recovery crew—some with smiles and some with tears—waved to us from the runway. Still wearing my heavy suit, I climbed slowly out of my seat, descended the ladder to the mid-deck, and gingerly walked across the bridge into the Crew Transport Vehicle. As soon as I entered the CTV, I began feeling light-headed and weak. I told the medical team, "I'm definitely not my normal self. I've never felt this way after previous missions."

The suit techs helped me out of my pressure suit. I headed into the bathroom, which was slightly larger than the restroom on an airplane. Although I wasn't sick, I felt like I would faint at any moment. I drank four cans of Dr Pepper, hoping the sugar and liquid would pick me up.

The rest of the crew was feeling fine, but their commander wasn't sure she could even walk down to the runway. We stayed in the CTV about fifteen minutes longer than usual, because I felt I might not be able to stand on my own. Finally, I mustered the strength to walk down the stairs to the runway.

I was surprised to see Joe Lanni, the wing commander at Edwards, who was my husband's roommate at the academy. I shook hands with the generals and other VIPs as we walked around to inspect *Discovery*. Although the fresh air and physical movement helped, every time I clasped someone's hand, I joked, "Please don't let go! Just hold onto my hand for a little while."

I wanted to check the tires after my fast landing. To my relief, they were in perfect shape.

Whew.

We loaded into a van and drove to the bachelor officer quarters. I took a thirty-minute shower and finally started feeling better. Two of the nurses attending to us, Katie Kirk and Vickie Johnson, were old friends of mine from Edwards. I hadn't seen them in fifteen years. They took my blood pressure while I chatted with them and wolfed down three sandwiches. My blood pressure was abnormally low, and my heart was racing.

We held a press conference, which I was honestly not in any physical shape to attend. I looked worn and tired—probably not as bad as I felt. However, it was important for us to address the huge media presence for this first landing after the *Columbia* accident.

Shortly after the press conference, we boarded the Shuttle Training Aircraft to return to Houston. I sat on the right side of the plane, staring out the window. I may have dozed off a few times. I didn't say a word to anyone the entire flight home.

Pat and the kids met me at Ellington Field. They'd been out at Kennedy, hoping we would land there. I had seen my kids only once in the last five weeks. Our crew said a few informal remarks to our friends, training team, and media. It was already after sunset. I'd been awake most of the past twenty-four hours.

The next day, I had my medical checkup at JSC, returned to Ellington for the public "crew return ceremony," played with my kids, sorted through the mail, and cleaned up around the house.

The following morning, my feet hurt so bad that I was unable to walk. I couldn't go to work. I just had to sit. This had never happened to me before, and I'd never heard other astronauts complain about it. I suppose it was from not using my feet for two weeks. I didn't put any pressure on my feet at all during the mission except for thirty minutes a day on the bicycle ergometer. That wasn't enough to keep them in condition.

I believe the bones in my feet changed position in zero gravity. When I came back to Earth, the bones had to move back into place again, pushing on muscles and stretching tendons. It took two weeks for the pain to eventually subside.

On my first space shuttle mission, I was thirty-eight years old. I was forty-eight during this final mission. It had been six years since my last trip into space, and that was only a five-day flight. This one was fifteen days. Even though I had exercised almost every day in space, I felt awful. My body was simply not what it used to be.

I believe that as you grow older, it becomes easier for you to go *up* into space and adapt to living there, but it is much harder to come *down* again. I initially blamed feeling worn and weak on my age. But other factors could include the intense nature of the flight, lack of sleep, dehydration, or the fact that I ate only about half of my packed food.

We have so much yet to learn about how the human body adapts to living in space. I was certainly fascinated by how differently I readapted each time I came home. We need to continue to collect data on long-duration stays such as those on the ISS—both the raw medical data and the subjective feelings experienced by astronauts.

Endings

We accomplished all of the major mission objectives. However, we couldn't accurately call this mission "an unqualified success." I reminded people that it was a test flight. Much had gone very well; we had proven new techniques and new technologies. But we had undoubtedly dodged a bullet with our external tank. The space shuttle would be grounded indefinitely until the problem could be fixed.

I was extremely proud of my crew's performance, with so many eyes watching us. Each member of the crew carried out their roles superbly. Our training team had been outstanding. I'm eternally grateful to the engineers and flight controllers and technicians who supported us all along the way. Everyone at NASA and its contractors gave this mission their best shot, no doubt about it. And yet, there was still so much work to be done.

Chaos reigned in my life much of the next year. Delta Air Lines declared bankruptcy in September, and Pat's retirement plan evaporated. After a long period of declining health, my mother passed away in Elmira on November 18, 2005—two days before I was scheduled to come home to visit her. That same month, my kids turned ten and five years old. In the midst of all of this, I was constantly giving interviews or flying around the country for public appearances.

I told Kent Rominger that I intended to retire on March 1, 2006. NASA wrote a press release, but I asked Kent not to issue it until closer to my retirement date.

In February 2006, I was finally able to take my celebratory post-mission trip home to Elmira to give presentations to schools and civic groups. My father drove down from Rochester to see me that week. I spent time with him every day, and we enjoyed catching up with each other. He told me that he'd just seen the doctor, who said he was "good for another ten years."

After I delivered talks at Edison and Notre Dame high schools on February 27, I said goodbye to Dad and flew back to Houston. Dad told the Elmira *Star-Gazette* that afternoon, "I'm in awe of her. I really am."

I returned home that evening to a phone call from my brother Jay. He said that at seven-thirty, a car struck my father as he started to walk across West Washington Avenue. Dad died from his injuries later that evening.

I had lost both my parents in a span of three months. As shocking and sad as it was, as I look back I am thankful they were able to experience my four space missions with me, with all their adventures and challenges.

The next morning, I called Kent and told him I needed to delay my retirement by two months. I needed to deal with my father's will and all that an adult typically faces after losing a parent. I finally retired from NASA and the astronaut corps in mid-May.

I could have changed my mind and remained an astronaut. However, I was keenly aware that when I came back from my fourth flight, there were fifty qualified astronauts who had not yet flown a single mission. There were only eighteen flights left on the manifest at that time for the remainder of the Space Shuttle Program.

Some of my peers stayed on for a fifth (or even a sixth or seventh) flight. You could make a valid argument that the program benefited from economies of scale when astronauts flew multiple times. You saved money by not having to retrain people for each mission.

In my mind, though, there were far greater benefits to the nation and the world if we flew those fifty astronauts at least once. Flown astronauts have a unique perspective that is difficult to replicate any other way. Even though more than five hundred people have flown in space, that is a minuscule number relative to the entire population. By flying our rookie astronauts, our society would gain an additional fifty individuals who could talk to the public, work in private industry, or use their expertise in other areas within NASA. The benefits from that investment would pay back many times over the cost of training those individuals.

I am proud that every astronaut who stayed on flying status was eventually able to fly at least one mission.

Returning from Oz

As I look back over my career and all the speed bumps I had to drive over, all the stop signs, red lights, flat tires, and dead batteries, I'm amazed I made it to the pinnacle of my dreams. I flew in space—not just once, but four times. I visited the International Space Station, the final goal of my astronaut career. I was able to experience what it felt like to float in microgravity and see Earth as a *planet*. I carried the responsibility of returning the space shuttle to flight after a terrible accident.

I find it crazy that the first woman pilot and commander was *me*. How in the world did that happen? As a young girl with no special talents, someone who stuttered and was painfully shy, I empowered my aspirations to become real through my decisions and work ethic.

Thinking about the period between my high school graduation and my last flight, I feel like Dorothy from the *Wizard of Oz*, getting her wish of traveling somewhere over the rainbow but then simply yearning to go home. I wanted to explore, experience new adventures, see magical places. Along the way there were challenges, surprises, and dangers. But in the end, after years of striving to travel in space, there was the paradox: *on the last day of my last mission, all I wanted was to go home to be with my family.*

After landing from STS-114, with so much change and turmoil in my life, it was hard to focus. I felt I had to live just one day at a time. I empathized with other astronauts who had to deal with the fact that it was time to move on. My parents had always reminded me to recite the "Serenity Prayer" to draw serenity, acceptance, courage, and wisdom.

My parents instilled in me the importance of faith in God. Throughout the challenging moments of my missions and the other difficult times in my life when I needed strength, prayer provided me with courage, confidence, focus, and a knowledge that there is something greater than myself.

I spent forty years of my life in pursuit of flying farther, faster, and higher. I accomplished so much of what I dreamed of, more than I imagined was possible. And now, it was time to devote myself to serving my family, my community, and my country by helping to inspire the leaders of tomorrow.

AFTERWORD

ON MISSION

When I asked myself *why* I wanted to write this book, I came up with this list:

- To add to the historical record of the Space Shuttle Program
- To document the events in my life, including my experience on my four space missions
- To inspire young people to choose careers in the military
- To inspire young people to choose careers in the space program
- To share the lessons I learned about how to be an effective colleague, leader, and trusted ally
- To raise awareness of space issues
- To educate and inspire today's leaders to invest in space exploration
- To satisfy all those who ask me: "Where is your book?"

There is truth in all of these reasons, but my primary purpose was: *To inspire my readers to reach for the stars—whatever your "stars" may be.*

I have lived a very unusual and exciting life, and I have been blessed with an incredible career.

But why *me*? Why was *I* able to experience and achieve so much?

It certainly wasn't because of money or social status. I definitely had no outstanding athletic, intellectual, or other special abilities. I grew up as an unremarkable lower-middle-class American. Nothing in my early years would possibly have led anyone to guess what I would be doing a few decades later.

Being the first to do something by definition implies that the timing has to be right. But merely being in the right place at the right time isn't enough if you don't also recognize and take advantage of the opportunity.

There have been many people who were the first to do something but burned many bridges or stepped on many toes to make it happen. That's not who I am. I like to believe that my story shows that people who play by the rules and act with character and integrity can indeed finish first.

First and foremost, I would like to be remembered as a trustworthy person, because honesty and straightforwardness are qualities I value. I have always wanted people to know that I will be straight with them. Without trust, there's no basis for a relationship with someone whether at work or socially.

Intertwined with that, I would like to be remembered as being knowledgeable. As you've seen throughout my life story, I always made it a point to learn as much as there was to know about a subject I was entrusted to master. I studied, memorized, practiced, and learned by experience.

Another facet of knowledge is admitting to blind spots—knowing what you *don't* know. I've always tried to be someone who had the wisdom to say, "I don't know," and the humility to listen to others who might have better ideas or information than I.

Trustworthiness and knowledge together will make you a strong and reliable boss, employee, teammate, or friend. People can count on you for sound advice or to perform when you're entrusted with a task.

My imagination and visualization helped me identify possibilities and see the paths that might take me where I wanted to go. I can trace so much of this to my love of reading ever since I was young. It's a shame that reading seems to be a lost art in the world of today's social media distractions and "short-term memory devices." Reading gave me the impetus to strive for excellence and heightened my curiosity. I particularly valued stories about how people struggled to ultimately achieve success. There are so many important lessons to be learned from heroes in every imaginable domain. Crucial books I read in my youth told the stories of Thomas Edison, Douglas Mac-Arthur, Ernie Davis, Michael Collins, Galileo Galilei, Jackie Cochran, and many others. These were all examples of people with a strong passion to achieve something that truly mattered to them.

If passion lights the fire that gets you going, discipline and focus will keep you moving forward despite the inevitable setbacks along the way. Nothing ever goes perfectly according to plan. There are just too many variables in life to keep them all under control. And it's often true that the more important the goal, the more resistance we encounter in striving to achieve it. We frequently get in our own way, too. I made many boneheaded mistakes along the way! But I believe mistakes and setbacks are an important part of any meaningful journey—that's why I admitted to so many of them in this book. I even developed a four-step process for dealing with mistakes: admitting your mistakes, correcting them, making changes to avoid repeating them, and then moving on. Show up at the important work every day. Don't give up. Strive for what matters most.

Having a clear mission focuses all of these qualities. Whether the mission is for your organization, your family, your relationship, or even yourself, a mission statement will make explicit what you want to achieve, why it's important, and what the result will look like.

I am happiest and most engaged when I focus on goals, problem solving, helping others, and staying mission-oriented. In particular, I feel most fulfilled when I don't think about "what's in it for me." Service to others is central.

It's no coincidence that most of this book focuses on my military and space missions, where I was able to contribute to a much larger effort that benefited my country and humankind. All participants in the space program know they are part of something bigger: the human need to explore the unknown and the ultimate need to expand human presence beyond the surface of the Earth and into outer space.

We live in the greatest country in the world. I firmly believe that anyone here can strive for the kind of life they want to live. We can dream big and, if we're willing to do the work, make those dreams come true. I've seen many other countries where poverty and discrimination against gender or ethnicity prevent people from having the same kinds of opportunities I had. This is a terrible shame, but I believe it is slowly changing.

It gives me all the more reason to remind people that we cannot take our freedom for granted. That's why patriotism is one of my core beliefs. It started for me with reading stories about men and women who took brave actions for a cause and was reinforced as I saw more of the world during my Air Force career. Whether defending freedom in the world wars, inventing devices to improve our lives, exploring new worlds, or presenting new theories that went against prevailing beliefs, these types of leaders have inspired and motivated me to carry on.

And while we're striving for great things, it's important to enjoy the journey along the way. We can't take ourselves too seriously. Having a sense of humor and knowing when to lighten the mood can make a huge difference in motivating ourselves and others. We must be able to laugh at ourselves, laugh with our friends and coworkers, and support one another with smiles whenever possible.

Finally, I wrote this book to stop that pesky question I've heard so many times since 1995: "Where is your book?" Well, here it is! I hope you've enjoyed reading it, and I hope I've been able to enlighten you with the excitement I felt as a leading participant in the exploration of the Earth and space.

I encourage you to continue reading stories of exploration and adventure. Remember that you too can go farther, faster, and higher, just as I have. The world needs more people—women and men—to break through the glass ceiling.

And to the stars? Well, as far as we humans know, the size of the universe is infinite, so you better get moving!

ACKNOWLEDGMENTS

Writing a book is a huge undertaking, involving research, interviews, deadlines, an enormous time commitment, and the risk of failure. After my final shuttle flight, I prioritized raising my two children. I continued to work, but with a more varied schedule. The idea of writing a book was simply not a realistic option, as my life was still progressing on fast-forward.

And so, fifteen years passed quickly by.

I have never been a professional author, so I continued to resist writing a memoir despite knowing that it was the right thing to do. The tipping point came when my coauthor, Jonathan Ward, contacted me just prior to the COVID-19 pandemic. He convinced me the time was right for this book, and I knew he was the missing link in getting this story completed. My thanks to him for his perfect timing. His willingness to work with me and his research, writing, and industry experience were foundational. It would not have happened without him.

Many other people played crucial roles to help us see this project through. Our spouses Pat Youngs and Jane Ward deserve praise for their loving support and patience with us during the many months of interviewing, writing, editing, and rewriting.

We are grateful to the special people who provided their time and thoughts for interviews and helped refresh my memory of eventful times: my brother, Jim "Jay" Collins, and my sister, Margy Conklin;

my daughter, Bridget Youngs; my cousin Mary Kay Morin; my Air Force colleagues Bob Vosburgh and Gordie Neff; astronauts Tom Jones, Jim Wetherbee, Charlie Precourt, Steve Hawley, Cady Coleman, and Wendy Lawrence; Wayne Hale and Lisa Reed from Johnson Space Center; and Mike Leinbach from Kennedy Space Center.

Jonathan's sister, Penny Ward, transcribed endless hours of our recorded conversations. We are grateful to those who reviewed and commented on the drafts of the book or chapters, including Pat, Bridget, and Luke Youngs, Mary Kay Morin, Francis French, Jane Ward, Tyson Bowser, Janice Codispoti, Janet Oliver, and Larry Puzio.

Thanks to our literary agent, Jim Hornfischer (who tragically passed away before the book went to press), and our editor, Cal Barksdale, for believing in this project and enabling us to bring this book to the public.

ABOUT THE AUTHORS

Colonel Eileen M. Collins, USAF (retired), earned a place in history as the first American woman to pilot, and later to command, a space mission.

Collins's love of aviation began as a child in her hometown of Elmira, New York, which hosts the National Soaring Museum. At age twenty, she used money saved from part-time jobs to take flying lessons. After earning an associate's degree in mathematics/science from Corning Community College in 1976 and a bachelor's degree in mathematics/economics from Syracuse University in 1978, she was selected as one of the first four women admitted to the Undergraduate Pilot Training Program at Vance Air Force Base, Oklahoma. In 1979, she became the first woman flight instructor at Vance. As a C-141 Starlifter transport pilot, Collins participated in the 1983 invasion of Grenada, delivering troops and evacuating medical students.

Collins earned a master's degree in operations research from Stanford University in 1986 and a master's degree in space systems management from Webster University in 1989. Following a three-year stint teaching mathematics at the Air Force Academy, she graduated from the Air Force Test Pilot School at Edwards Air Force Base, California, in 1990, the second woman pilot to do so.

Collins became a NASA astronaut in July 1990. She flew as the STS-63 pilot in 1995, aboard *Discovery*, the first shuttle to rendezvous with the Russian space station Mir. She was also the pilot for STS-84 on *Atlantis* in 1997, when her crew docked with Mir. Collins became the first woman commander of a US spacecraft with shuttle mission STS-93 on *Columbia* in 1999, deploying the Chandra X-Ray Observatory. Her final spaceflight was as commander of STS-114 on *Discovery* in 2005, NASA's "return to flight" mission after the tragic *Columbia* accident.

When she retired from the Air Force in January 2005 and from NASA in May 2006, after a distinguished twenty-eight-year career, she had logged more than 6,751 hours in thirty different types of aircraft and had spent 872 hours in space.

Collins has served on numerous boards and advisory committees, including the NASA Advisory Council, the United Services Automobile Association, the National Space Council Advisory Group, the National Academy of Science's Aerospace Science and Engineering Board, and the Astronaut Memorial Foundation. She was inducted into the National Women's Hall of Fame in 1995, the National Aviation Hall of Fame in 2009, the Astronaut Hall of Fame in 2013, and the Texas Aviation Hall of Fame in 2020.

Among her many awards and commendations are:

- Defense Superior Service Medal
- Distinguished Flying Cross
- Defense Meritorious Service Medal
- Air Force Meritorious Service Medal with one oak leaf cluster
- Air Force Commendation Medal with one oak leaf cluster
- Armed Forces Expeditionary Medal for service in Grenada (Operation Urgent Fury, October 1983)
- French Legion of Honor
- NASA Outstanding Leadership Medals
- NASA Space Flight Medals
- Harmon Trophy
- Free Spirit Award
- National Space Trophy

Collins is also a member of the Air Force Association, the Order of Daedalians, Women Military Aviators, Women in Aviation International, the American Institute of Aeronautics and Astronautics, and the Ninety-Nines. She is represented by Keppler Speakers.

Collins is married to retired USAF and Delta Air Lines pilot Pat Youngs.

Jonathan H. Ward is an author, Fellow of the Royal Astronomical Society, Solar System Ambassador for the Jet Propulsion Laboratory, and a frequent speaker on space exploration topics to astronomy clubs, interest groups, and at regional and national conferences.

His love of sharing the thrill of space exploration with the public began in high school, when he served as a volunteer tour guide at the Smithsonian's National Air and Space Museum during the Apollo 15 and Apollo 16 missions. He worked for Boeing on the Space Station Freedom Program in the late 1980s.

After studying physics at Carnegie-Mellon University, Ward transferred to Virginia Commonwealth University, from which he graduated summa cum laude in 1978 with a BS in psychology. He received a Master of Science in systems management from the University of Denver in 1992. Ward is credentialed as a Professional Certified Coach by the International Coach Federation and serves on the adjunct staff at the Center for Creative Leadership.

As an author, Ward is a passionate defender of the perspectives of others. His insight enables him to relate without embellishment the many truths and motivations at play in complex situations. He enjoys telling stories of unsung heroes—everyday people who find themselves in the midst of extraordinary circumstances and who rise above adversity to achieve success against seemingly impossible odds.

His 2018 book, *Bringing Columbia Home: The Untold Story of a Lost Space Shuttle and Her Crew*, coauthored with former space shuttle launch director Michael Leinbach, earned accolades ranging from a starred review in *Kirkus Reviews* to features on CBS News and in *USA Today*, *Salon*, and the *Christian Science Monitor*. His first book, *Rocket Ranch: The Nuts and Bolts of the Apollo Moon Program at Kennedy Space Center*, was the bestselling engineering title of 2015 for academic publisher Springer/Praxis—quite an accomplishment for a nonengineer author.

Ward is a former president and section leader of the Washington Chorus, with whom he performed at Carnegie Hall, as a soloist at the John F. Kennedy Center for the Performing Arts, and on European tours. He sang on two GRAMMY®-winning albums (John Corigliano, *Of Rage and Remembrance*, National Symphony Orchestra, 1996; and Benjamin Britten, *War Requiem*, the Washington Chorus and Orchestra, 2000).

Ward spent several years of his childhood in Japan and considers the Virginia suburbs of Washington, DC, to be his hometown. He and his wife, Jane, reside in Greensboro, North Carolina.

INDEX